NAVIGATION CONSTELLATION DESIGN
USING A MULTI-OBJECTIVE GENETIC ALGORITHM

Chapter 1
Introduction

In order to maintain the power and influence the United States (US) currently possesses in space, innovation and the will to try new ideas is imperative. In today's world, money and budgets are a priority when it comes to analyzing space systems. Attempting to maintain or improve performance but decrease the cost is the goal for many military and civilian leaders. The Global Positioning System (GPS) is a critical system for the military and civilian sectors of the US. Positioning and timing for systems such as precision guided munitions, banking, power grids, and basic navigation are provided by GPS. The US Air Force is concerned with researching new configurations for the GPS constellation to continue to provide accurate position and timing measurements for global use. Analyzing the tradeoffs between performance and cost, the following research determines there are alternate navigation constellation designs at various altitudes.

1.1 Motivation

To improve coverage capabilities provided by single satellites, satellite constellations are used. For missions, such as GPS, where constant global coverage is a requirement, constellations are a necessity. Satellite constellations also add robustness to a system because if one fails, there are still multiple satellites functioning properly. Over the years, the need for global coverage has increased, and navigation is not the only mission seeking to utilize constellations for this purpose. Missions like communications and remote sensing also require the benefits that constellations offer.

Since the need for satellite constellations has increased through the years, there has been more focus on the design process for such systems. However, designing a satellite constellation has many different segments that must come together properly in order to achieve the mission. There is the orbit design and configuration segment, which determines the amount of coverage the constellation will provide. Spacecraft design is the segment which relies on the altitude and determines spacecraft cost. The launch manifest segment relies on altitude, but also relies on other parameters such as satellite inclination and the number of planes in the constellation. Lastly, cost through deployment takes into account the research and development, production, and launch costs of the system. Figure 1-1 illustrates the multiple factors that have an influence on the design of a constellation [1].

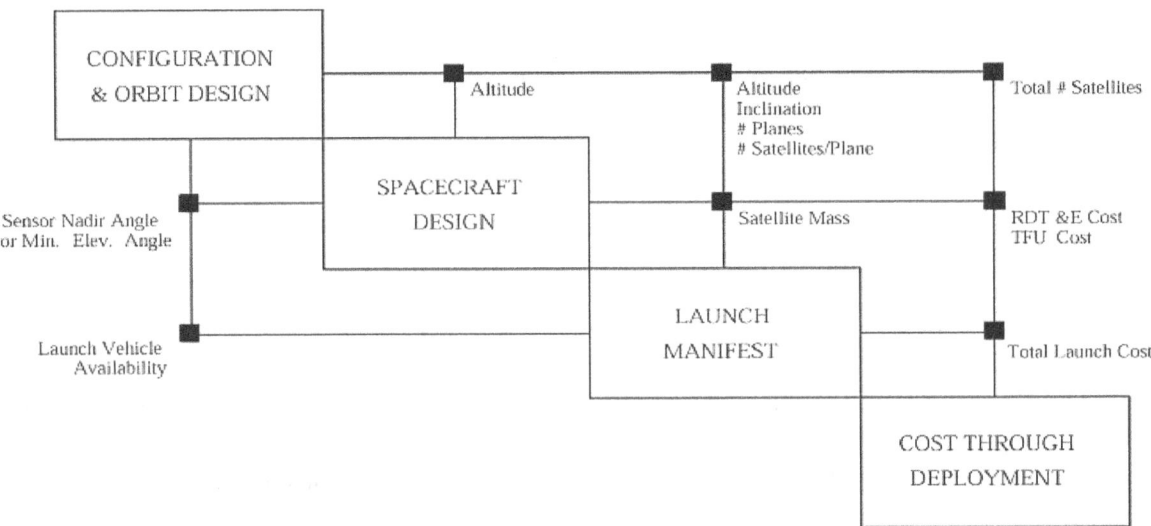

Figure 1-1: Satellite Constellation Design Problem

The task of determining a specific constellation design forces decision-makers to evaluate possible solutions against the mission requirements. When requirements of a system conflict, there are tradeoffs. The tradeoffs are used to determine the best solution based on what is most important to the decision maker. As a result of the complexity in the design process, constellation design tools are used to determine possible solutions. When analyzing tradeoffs of a system, the multi-objective genetic algorithm (MOGA) is well suited for the problem [2]. MOGAs approximate a Pareto front, which is a set of designs to a multi-objective optimization problem. MOGAs are most beneficial when the objective functions in the problem are conflicting. Performance and cost are usually conflicting factors in a system. MOGA's are able to consider multiple design variables and their relation to the objective functions in the problem [2].

The GPS constellation is an example of a system that used constellation design tools to analyze the potential performance and cost of the program. The GPS was developed by the Department of Defense (DoD) in 1973, and it was initially intended to only be used for military needs. As a result of the wide range of uses provided by this system, the DoD later promoted civilian use of the system [3]. The GPS consists of three segments: the space segment, the control segment, and the user segment. The space segment is required to maintain the availability of at least 24 satellites, so currently the GPS constellation possesses 31 operational satellites that orbit at an altitude of approximately 20,000 kilometers (km) [4]. Each satellite circles the Earth twice a day. The constellation is currently designed with six equally spaced orbital planes, and each plane holds four satellites [4]. The 24 satellite arrangement guarantees that a user on the Earth can view at least four satellites from any point on the planet. The need for GPS has grown tremendously over the years, which is the reason the US has continuously given money to the research and development of the program. GPS is considered a key piece to mission success, for missions involving navigation, in the US due to its precise positioning and timing capabilities.

> "Efficiencies in positioning, movement and timing derived from the ubiquitous GPS signals have already quietly permeated virtually every level of our national infrastructure to the extent that, in many cases, there is no going back to earlier ways of doing things without tremendous but unrecognized penalties" [5].

Ensuring that US capabilities are maintained is a crucial task for both military and civilian decision makers. It is fundamental that we research different constellation design tools that are focused on navigation systems so that navigation constellation designs can be analyzed and compared.

1.2 Problem Statement

As the technology on the satellites has improved throughout the years, the cost to develop and produce the satellites has increased. In 2013, the House Armed Services Committee directed the Air Force to report on lower-cost GPS solutions [6]. Combining the research conducted on the current GPS constellation with the capabilities of a constellation design tool, it is possible to develop new designs for navigation, designs with similar performance levels to that of the current GPS but with lower costs.

1.2.1 Design Variables.

This research analyzes navigation constellation designs using a constellation design tool that consists of MATLAB's MOGA and Analytical Graphics, Inc. (AGI) Systems Tool Kit (STK). This research uses Walker constellations because in this constellation, all the satellites are equally distributed on similar and phased orbital planes. In order to describe the constellation and satellite design, design variables, separated into three main categories, are included in this research. The first category is the Walker parameters, which include the number of planes in the constellation, the number of satellites per plane, the true anomaly phasing, and the right ascension of the ascending node increment. Depending on the altitude, these parameters will vary in order to maintain global coverage. The second category of variables is the orbital parameters:

> alt: altitude of the orbit in km
> e: eccentricity of the orbit
> i: inclination of the orbit in degrees
> ω: argument of perigee in degrees
> Ω: right ascension of the ascending node in degrees
> M: mean anomaly in degrees

Section 2.1.2 discusses these variables in further detail. Each of these variables defines different aspects of the orbits included in the constellation. These parameters will vary to analyze the design changes for constellations at different altitudes. The last category of variables is the satellite design parameter. In this thesis, transmit power, measured in Watts (W), is the variable that determines the size of the satellite. With a higher altitude, more transmit power is required to close link at the edges of the Earth, so the size of the satellite will also be larger to accommodate the larger equipment needed. This variable constrains the size of the satellite used in specific scenarios to maintain a level of reality in the analysis.

1.2.2 Objective Functions.

The constellation design tool developed in this research possesses multiple objective functions. In constellation design, it is very difficult to optimize one objective function without considering other factors. By using a MOGA, tradeoffs between objective functions can be analyzed. The first objective function in this tool is cost, which is calculated by standard cost models from *The Space Mission Engineering: The New SMAD* [7]. The cost model takes into account the research and development, production, and launch cost. This cost model is just an estimation used in this tool to give users a general idea of the cost required for a given system. Due to the manner in which this tool is designed, a more accurate cost model, if available, can be used instead. However, there are limitations to the cost models used in this tool due to the assumptions made for this research. These will be discussed more in depth in later sections.

The second objective function in this tool is the position dilution of precision (PDOP) function. This function measures the performance of the navigation constellations. The position accuracy depends on the geometry of the satellites overhead. Navigation constellations can be compared using PDOP. Smaller values of PDOP are more likely to provide accurate positioning to the user. STK is used in this tool to perform the PDOP calculations, and it uses a basic line of sight concept to complete the calculations.

In order to minimize two objective functions (cost and PDOP), MATLAB's MOGA tool is used in this constellation design tool. The main goal of a MOGA is to determine all possible tradeoffs among the conflicting objective functions. Since tradeoffs between the functions are the outcome, it is difficult to obtain one single design without iterative interaction with the decision maker. The approach used in this research is to show a set of Pareto optimal solutions to the decision maker. This allows the decision maker to select one of the Pareto optimal solutions based on broader considerations not incorporated into the optimization. A Pareto front of solutions also illustrates guidelines of good systems.

This research combines the optimization of multiple objective functions using MATLAB's MOGA, the cost model written in MATLAB, and STK's calculation of PDOP to create a navigation constellation design tool. Using the design variables described above along with the objective functions, this research analyzes alternate navigation constellation designs at different orbital altitudes and evaluates the trade-offs of the various designs. Using the optimized constellations, different design guidelines are articulated.

1.3 Assumptions

To define the scope of the research, assumptions were made throughout the process. Each objective function was created separately and written in individual MATLAB files. The cost function developed in this research is assumed to be a sufficient estimate for the scenarios analyzed in this thesis. It is a cost model developed specifically for a scenario using Walker constellation designs. Because a Walker design is used, it is assumed that every satellite in the constellation is exactly the same. Therefore, in the cost model, the cost is determined for one satellite and then multiplied by the total number of satellites in the scenario. The goal of this research is not to produce a cost model; it is to use a cost estimate to analyze multiple navigation constellation designs.

Within the cost function, there is a calculation to determine the type and number of launch vehicles needed for the problem. The process of fully optimizing the number and specific type of launch vehicles used in a scenario is outside the scope of this research due to it being a combinatorial problem. Therefore, the calculation to determine launch vehicles used in this work is assumed to be an adequate estimation of the vehicles required in a scenario.

To analyze performance, the PDOP function is used. The values of PDOP are calculated through STK, and the measurement is assumed to be an indicative metric. Using STK does result in weaknesses in the metric, but overall, it is assumed to be reasonable estimate when compared to the current GPS constellation values. Once the MOGA has produced a Pareto front, it is assumed that the solutions on the front are optimal.

1.4 Research Objectives

This research consists of two main components: the development of the design tool and the analysis of optimal design solutions that are produced. The first component requires research into constellation design tools and the modification of the tools to operate for navigation satellites. The cost function, PDOP function, and overall MOGA function must operate together in order to produce design solutions. The functioning tool will be used to explore the design trade space of navigation constellations. By applying this method, it allows for a broader search than feasible by hand to be accomplished. The automation of this design tool also allows for high fidelity simulations to accurately capture and illustrate the tradeoffs. The resulting tradeoffs of the design solutions will give decision makers design guidelines for navigation constellations at different altitudes.

1.5 Summary

Satellite constellation design tools are continuing to improve over time, and they still maintain their value since military and civilian leaders are consistently looking for methods to decrease cost but improve performance on space systems. This thesis is supplementing the constellation design tool research completed by Second Lieutenant Evelyn Abbate[1] by illustrating the use of MATLAB's MOGA tool, STK's PDOP calculation, and the interface between the two programs. This specific tool not only offers high fidelity simulations, but it also offers the visual aid provided by STK. The ability to demonstrate the design tradeoffs and illustrate solutions in STK gives decision makers the freedom to select solutions based on their broader considerations and their specific preferences. It is difficult to produce a complete design solution when only analyzing one objective function. This methodology incorporates the necessity of analyzing multiple objective functions and providing tradeoffs for navigation constellation designs.

---

[1] 2dLt Abbate was a graduate from the Air Force Institute of Technology (AFIT).

Chapter 2
Background

This chapter presents a summary of topics relevant to constellation design and optimization methods. It describes different constellation configurations that have been considered and specific constellation design tools. Various optimizers, along with their advantages and disadvantages, are analyzed. Finally, a description of previous and current work is given.

2.1 Concepts Related to Constellation Design

Over the last several decades, space has grown from a strategic asset into tactical applications that support the warfighter. Space influences government, business, and culture. Satellites deliver television broadcasts, weather forecasts, and navigation through GPS. The US has become heavily reliant on the use of space systems not just for the everyday use, but for carrying out military operations. Each satellite has a purpose, and the orbit it is placed in determines how that purpose is carried out. According to James Wertz, an orbit is defined as the path of a spacecraft or natural body through space. More specifically, a Keplerian orbit is "one in which gravity is the only force, the central body is spherically symmetric; the central body's mass is much greater than that of the satellite; and the central body and satellite are the only two objects in the system" [7]. The following sections describe the different orbital altitudes, classical orbital elements, and perturbations that exist for different orbits.

2.1.1 Orbit Types.

There are different categories of orbits. It is important to understand these for the

purpose of this research. Orbital altitudes include three main categories: low Earth orbit

(LEO), mid Earth orbit (MEO), and geosynchronous orbit (GEO). The altitude is

measured from the surface of the Earth. A satellite in LEO is in an altitude range of 180-

2,000 kilometers (km). MEO is classified as any orbit between 2,000 and 35,780 km.

Lastly GEO is an orbit at 35,780 km [8]. Each orbital altitude provides different benefits

to a mission. For example, satellites in LEO can be used for imaging because they are

closer to the Earth's surface. However, GPS is required to have global coverage, so it

benefits from being in MEO and GEO. Spacecraft placed in GEO are used for Earth

observation and communications [9]. The significance of this orbit is that its period[2] is

equal to the rotation period of the Earth, which means its period is one day. This allows a

spacecraft to remain fixed over a point on the Earth's equator.

2.1.2 Astrodynamics.

To define the orbit of a spacecraft around the Earth, there are six parameters that

are necessary to recognize, and these parameters are referred to as classical orbital

elements (COEs) or Keplerian elements. These parameters include semi-major axis,

eccentricity, inclination, right ascension of the ascending node, argument of perigee, and

true anomaly (Figure 2-1). To describe an orbit's shape and size, the semi-major axis and

eccentricity are used. The semi-major axis measures the orbit size; it is half the length of

---

[2] The orbital period is the time it takes for a spacecraft to revolve once around its orbit.
[11]

the major axis of the ellipse. The important fact to note with this variable is that it is not measured from the surface of the Earth, so altitude and semi-major axis are different. Eccentricity is used to measure the shape of an orbit, and it is the ratio of the semi-minor to the semi-major axes. Table 2-1 shows the different conic sections with the associated values for the semi-major axis and eccentricity [7].

Conic	Semi-major Axis	Eccentricity
Circle	=radius (>0)	=0
Ellipse	>0	0<e<1
Parabola	∞	=1
Hyperbola	<0	>1

Table 2-1: Orbit Properties of Conic Sections

Once the orbit size and shape are determined, the next few variables describe the orientation of the orbit plane. Inclination is the angle between the orbit plane and a reference plane, as seen in Figure 2-1 [10]. When the inclination is between zero and 90 degrees, it is said to be a prograde orbit. When the inclination is between 90 and 180 degrees, it is a retrograde orbit. The intersection of the equatorial and orbital plane through the center of Earth is considered the line of nodes, which is shown in Figure 2-1. When the satellite passes the equator from south to north, it is called the ascending node, and when it passes from north to south, it is called the descending node. Along with the inclination, the orbit is defined with respect to the line of nodes. The right ascension of the ascending node (RAAN) is the angle measured eastward from the vernal equinox[3] to the ascending node of the orbit [7].

Next, it is important to define the variable associated with describing the orientation of the orbit within the plane. To describe the rotational orientation of the major axis, argument of perigee is used. This is measured as the angle between the ascending node and the direction of perigee in the orbit. Lastly, true anomaly is used to describe the position of the satellite within the orbit. It is measured from the direction of perigee to the actual location of the satellite in the orbit [7]. With all of these orbital parameters together, a full description of an orbit is provided, and the location of a satellite can be determined.

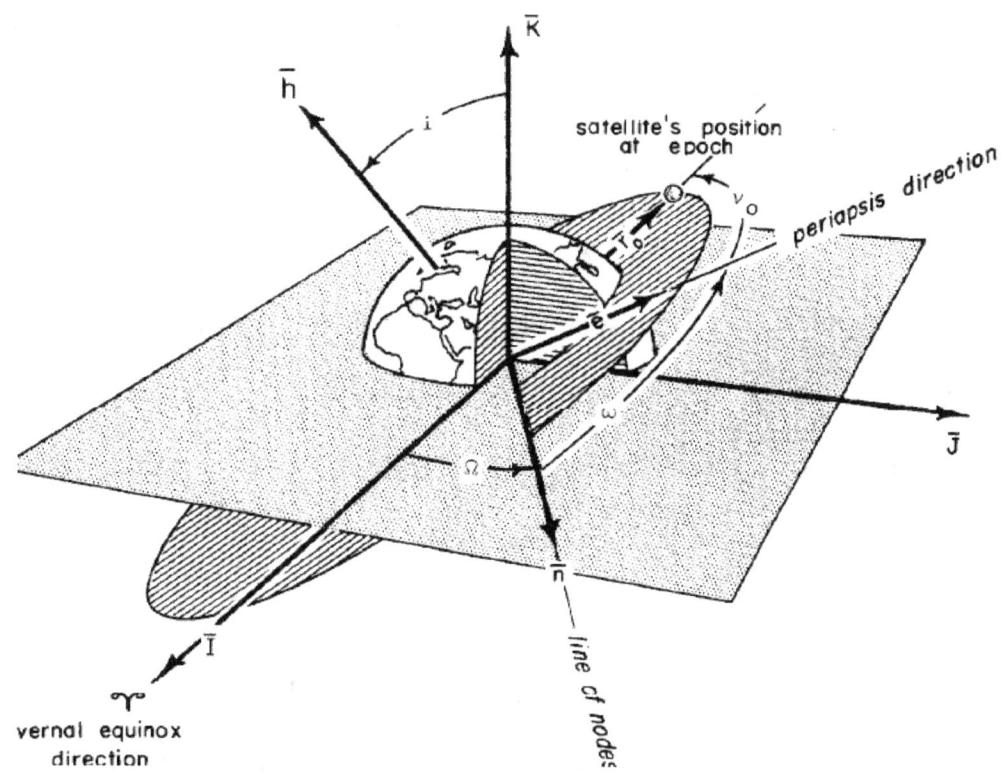

Figure 2-1: Classical Orbital Elements [4]

---

[3] Vernal equinox is the location of the Sun in the sky on the first day of spring. It is a reference point used for inertial space [7].

2.1.3 Perturbations.

The COE's described in Section 2.1.2 were defined using several different assumptions. The first assumption was that gravity was the only force acting on the spacecraft. Second, the Earth is spherically symmetric with uniform density, thus it could be treated as a point mass. Lastly, it was assumed the spacecraft mass remained constant. Real orbits do not follow these assumptions, but Keplerian orbits do give a reasonable estimate to a true orbit. A true orbit is normally defined for times in the past, and then using the satellite's ephemeris data, the orbit can be propagated into the future. When there is a change in any of the orbital elements due to outside forces, it is called a perturbation. Perturbations can have specific effects depending on the orbital altitude and positioning. Since perturbations vary at different altitudes, they can have an effect on constellation designs [11].

At lower altitudes, it is possible to still observe effects from the Earth's atmosphere. Gravity is not the only force acting on a satellite, and at orbital altitudes up to 600 km, atmospheric drag is a large force [11]. Drag affects the semi-major axis and eccentricity of the orbit because it removes energy from the orbit in the form of friction. One of the challenges with constellation designs at these altitudes is that drag is difficult to model because there are many factors related to it.

Another common perturbation comes from the Earth's oblateness. The Earth is not perfectly spherical, which perturbs the spacecraft because the gravitational force is not coming from the Earth's center. This perturbation is referred to as the J2 effect[4], and

---

[4] J2 is a constant describing the size of the bulge in the mathematical formulas used to model the oblate Earth [11].

it influences two of the orbital elements. This oblateness causes the orbit to precess, similar to a spinning top, which then affects the location of the ascending node. The rate at which the node changes is called the nodal regression rate, and it is a function of inclination and orbital altitude. Orbits at lower orbital altitudes and lower inclinations have the most influence from the J2 effect [11]. Along with the change in the ascending node, the J2 effect also influences the argument of perigee in an orbit. The effects are similar to the nodal regression rate in that the perigee rotation rate increases with lower orbital altitudes and lower inclinations. Orbit designs in LEO and MEO with lower inclinations need to take this perturbation into account [11].

Orbits in GEO have to consider the effects from solar radiation pressure. Sunlight consists of photons, which when in contact with a surface, transfers its momentum to that surface. When that small force is exerted unevenly on the surface of a spacecraft, it can cause slight movements over time. This poses a challenge especially for spacecraft needing precise pointing [11]. Each orbital altitude possesses its own challenges, but as long as the designer is aware of them, the mission can include mitigation methods. For this work, perturbations were not accounted for, but this section gives an idea of possible advantages and disadvantages of using certain orbits.

2.1.4 Constellation Types.

To improve certain mission objectives, a set of satellites, working to achieve the same goals, can be distributed over space. This is considered a satellite constellation. Since this research is focusing on global coverage, Walker constellations will be used in the design process. Walker constellations consist of circular orbits of equal altitude and inclination, and the orbital planes are spaced equally around the equatorial plane. These constellations are defined by the number of planes, the number of satellites per plane, and an inter-plane spacing. Their greatest advantage is that there are a finite number of them, and they can be identified and investigated [12]. Figure 2-2 illustrates the concept of a Walker constellation [13].

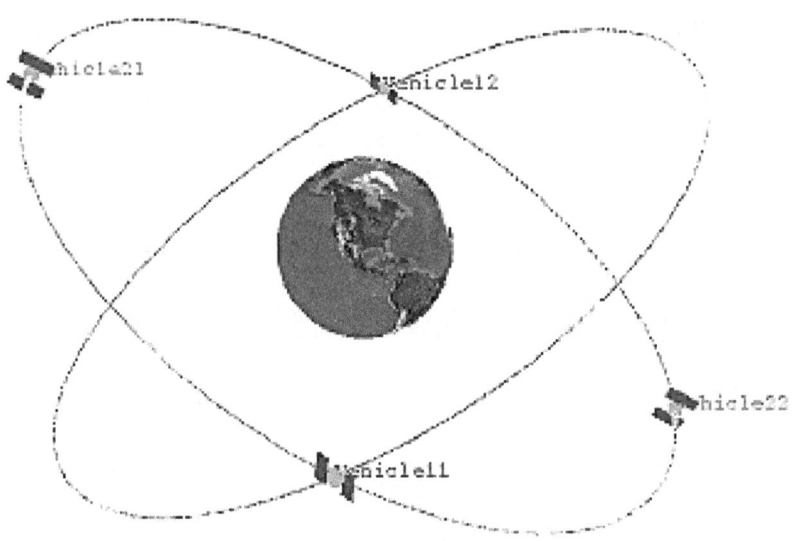

Figure 2-2: Walker Constellation

Although Walker constellations provide global coverage, there are several non-Walker constellations that can be used for regional coverage. Figure 2-3 illustrates several examples of non-Walker constellations [7]. Option A provides coverage over the polar region with less interest on the equatorial region. If coverage is needed in the equatorial region, along with the polar region, it is possible to add a plane over the equator. This is seen in Option B. Option C uses two perpendicular planes offset from the equator. Lastly, Option D allows for coverage of the polar region with better

2-16

coverage of the equatorial region compared to Option A [7]. The mission of the system will provide details on the type of constellation to utilize.

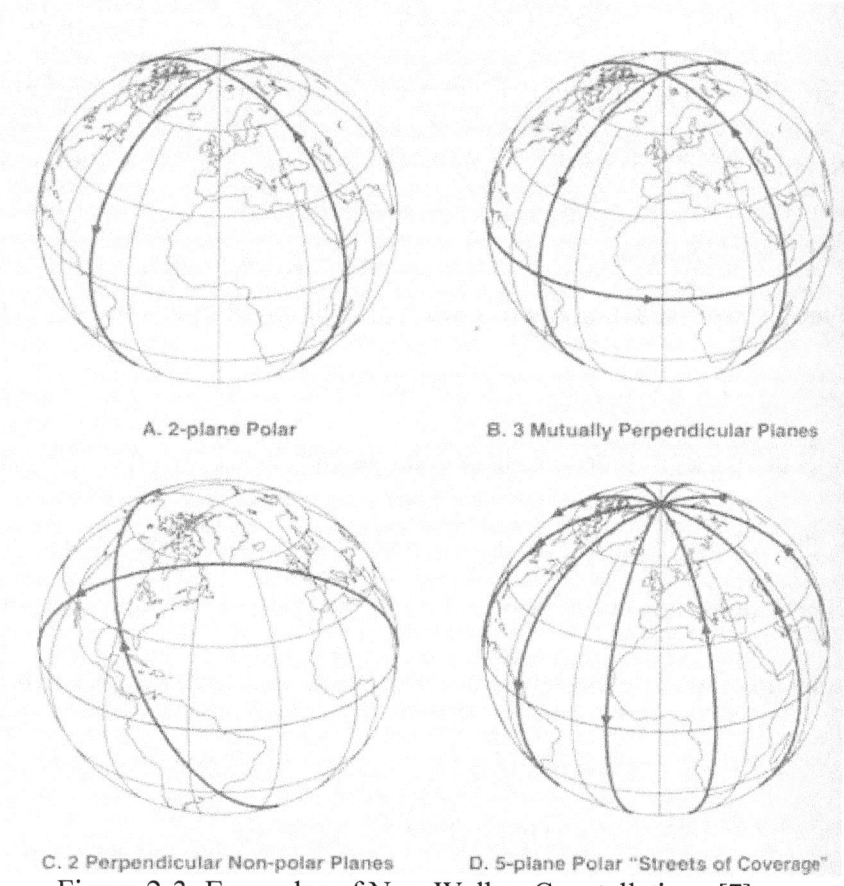

Figure 2-3: Examples of Non-Walker Constellations [7]

2.2 GPS Constellation Design

To facilitate the study of navigation constellation design, it is important to understand the design of the current GPS constellation. This section discusses various configurations for the GPS constellation in order to maximize coverage. Global coverage drives the design. To maintain global coverage, there is a need for good geometric diversity worldwide. As a result, the use of geosynchronous satellites is possible only with the proper amount of inclination.

The focus of navigation satellites started on MEO and LEO. However, the constraint on the design to maintain at least four satellites in view of the user at all times drove designs to higher altitudes. With the higher costs related to GEO and the poor geometric properties in LEO, MEO altitudes were considered the most beneficial for the GPS constellation [4]. The current constellation consists of near circular orbits with a radius of 20,200 km [14].

To meet the optimal amount of coverage, dilution of precision[5] characteristics, and cost, GPS satellites were placed in inclined 12-hour orbits, which are considered MEO. Orbits at higher altitudes produce good geometric properties and require a fewer number of satellites to maintain redundancy of coverage. Although the geometric properties and redundancy are improved, 12-hour orbits require station keeping, which means frequently correcting the satellites trajectory to stay in the correct orbit. To satisfy certain robustness considerations, multiple satellites were placed in specific orbital planes with an inclination of 55 degrees [14]. The total number of planes was chosen as six, with four satellites per plane. The planes are equally spaced by 60 degrees, but the satellites are not equally spaced within the planes [4]. This configuration is considered a tailored Walker constellation since the satellites are not equally spaced within the planes [4].

---

[5] Dilution of precision is a measure of the satellite geometry, which is referred to above as geometric properties

2.3 Constellation Design Tools

The decision to place satellites in a specific constellation includes factors, such as mission objectives, cost, available launch vehicles, and operational requirements to support the mission. There is no correct answer when it comes to constellation design, because each mission will require a different constellation configuration. According to Wertz, constellation selection is a process instead of set computations [7]. The following discussion provides a general process for designing a satellite constellation.

2.3.1 Constellation Design Process.

This section discusses the constellation design process described by Wertz [7]. One of the first steps is to establish the orbit type: Earth-referenced or space-referenced orbits. As the name implies, Earth-referenced orbits provide coverage of the surface of the Earth or near-Earth space. Since this research is focusing on navigation, the orbit type will be Earth-referenced orbit. Second, the orbit-related mission requirements need to be established. This includes factors like orbital limits and altitude needed for coverage. Requirements for resolution or launch capability constrain the orbit to lower altitudes; however, coverage, lifetime, and survivability drive the orbit to higher altitudes.

Satellites located at GEO maintain a fixed location over the Earth. The covered area of the Earth is a function of the minimum elevation angle of the specific system. Figure 2-4 illustrates the coverage from GEO as a function of the minimum elevation angle [7]. The highest percent coverage results with a minimum elevation of zero degrees, which means the system is viewing tangential to the Earth.

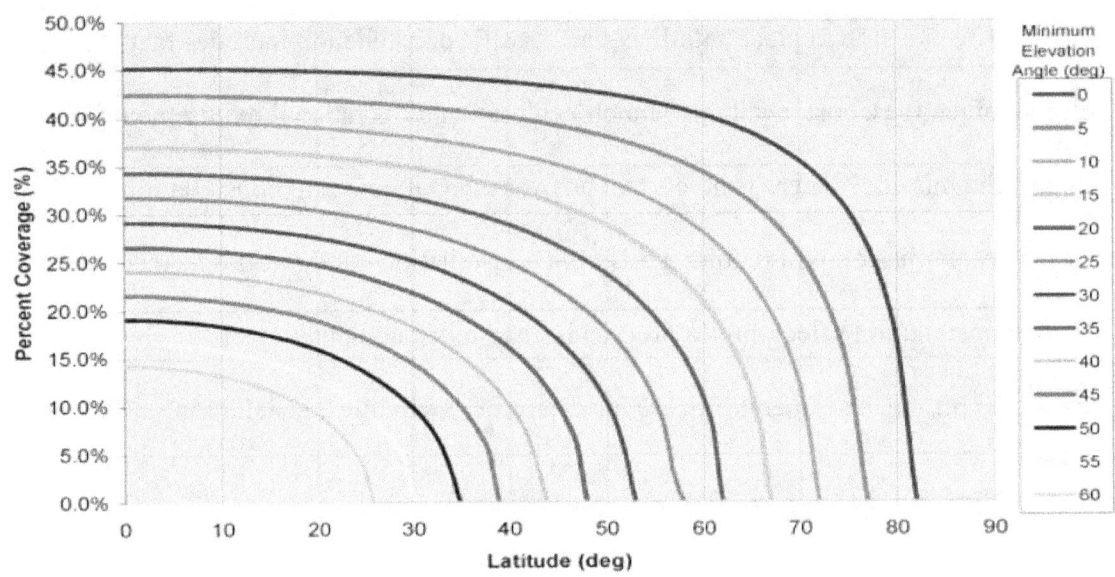

Figure 2-4: Coverage from Geostationary Orbit [7]

The third step in the design process is to evaluate the orbit. This step incorporates the decision of using a single satellite versus a constellation. Although a single satellite reduces mission cost, a constellation provides better coverage, higher reliability, and greater survivability. A constellation is also needed to create the necessary geometry for navigation and continuous coverage of the Earth. Fourth, the orbit cost must be analyzed. The previous step illustrates the orbit performance for a specific mission, but with greater performance usually comes higher costs. Launching a spacecraft into GEO requires launching approximately five times the spacecraft mass in LEO. Lastly, documentation and iteration is an imperative component of this process because it allows the design to be re-evaluated as mission conditions change over time [7].

2.3.2 Constellation Design Tools.

Designing a satellite constellation requires the consideration of several different variables and system requirements. In order to ensure the management and efficiency of the design process, different algorithms and tools have been developed. These tools are able to take in to account different aspects included in a constellation. For example, the altitude and inclination of a constellation have a large impact on the coverage performance, so the design tools are able to measure the optimal value of the two variables. Navigation is a well-known use for satellite constellations, so it has been used in validating different design tools over the years.

2.3.2.1 ORION.

Three specific satellite constellation design algorithms were studied and tested by GMV[6]. GMV integrated the algorithms with a software tool kit called ORION. This tool combines optimization procedures and Monte Carlo simulation techniques and gives the designers a method to plan and analyze a generic constellation. The algorithms were applied to the FUEGO constellation to validate their success. The FUEGO constellation is a constellation of small satellites that detect forest fires from LEO.

The first algorithm described is related to Walker constellations, where the traditional approach is to minimize the total number of satellites. It is considered the Symmetric, Inclined Constellation Design Method, and it optimizes the Earth central angle, θ. Figure 2-5 illustrates the geometry [15]. The user does not need to know all the details of the constellation to use this algorithm. However, with more parameters, the

---

[6] GMV is an international business group that operates in aeronautics, space, defense, and other areas of technology.

algorithm will produce more useful results in a timely manner. Some of the values to specify may include minimum elevation angle, minimum altitude, minimum and maximum number of satellites, minimum and maximum orbital plane inclination, and the number of orbital planes. The more efficient constellation is one with a lower value of θ and a fixed number of satellites [15].

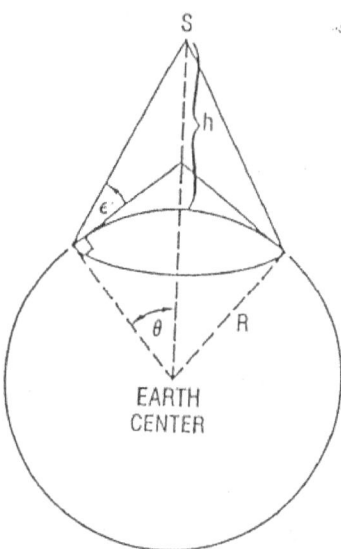

Figure 2-5: Single Satellite Viewing Geometry

The second algorithm is called the Polar, Non-Symmetric Design Algorithm, which uses a "streets-of-coverage" approach. This can be seen in Figure 2-3 Option D. This approach includes multiple circular orbit satellites at the same altitude and in a single plane. This creates a coverage band which is continuously viewed. Figure 2-6 illustrates this concept [15]. The only required input for this algorithm is the type of coverage to analyze, global or regional. Some additional parameters may include minimum elevation angle and/or altitude, number of satellites, desired folds of coverage, and number of orbital planes. The optimization methodology is implemented by using a series of analytical relations that determine the parameters that define the constellation.

2-22

The optimally phased polar constellations are determined by minimizing θ. This program outputs the optimal values of the angular spacing between orbits, the inter-plane spacing between satellites, and the value of θ [15].

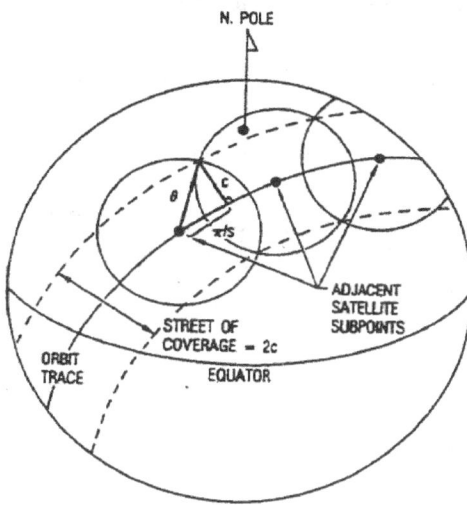

Figure 2-6: Street of Coverage from a Single Orbital Plane

The Advanced Adaptive Random Search Algorithm is the last algorithm analyzed in [15]. The two previous algorithms are commonly used in constellation design, but they have several limitations. They can only take classical geometric design factors into account, and they cannot take satellite failures or hybrid constellation possibilities into consideration. The Advanced Adaptive Random Search Algorithm is used to address those limitations. This is a variation of a genetic algorithm, but instead of optimizing multiple traits simultaneously, this algorithm optimizes each trait separate from one another. A trait is an aspect of the individual that influences the objective function. Then the optimized value of the first trait is used for the optimization of the remaining traits. In the optimization problem, there is a clearly defined function to optimize, and with this

2-23

algorithm, the following functions can be optimized: any DOP values, mean revisit time, maximum revisit time, and satellite failures [15].

Each of these algorithms is embedded in the constellation design tool. The first algorithm was used to define a Walker constellation based on required coverage. The second algorithm was used to define a polar, non-symmetric constellation based on required coverage. Then the last algorithm is used to optimize the selected constellation design that resulted from either of the other algorithms. These algorithms can be used separately or together, and if performed together, it creates a two-step optimization tool for constellation design.

2.3.2.2 Collaborative Optimization.

A new approach for system integration and optimization of a satellite constellation was studied in [1]. The conceptual design of a satellite constellation is a very complex, highly constrained, and multidisciplinary problem. Therefore the methodologies provide a way for designers to quickly analyze and explore design spaces [1].

The chosen methodology in [1] is collaborative optimization (CO), which was proposed by Braun [16]. The authors describe CO as an approach "that employs a bilevel optimization technique wherein a single system optimizer orchestrates and coordinates several optimization processes at the subspace level, that is, at the individual discipline or subsystem level." The authors suggest that CO implementation can benefit the constellation design problem.

The CO consists of a system level variable vector that possesses the top-level design parameters. Those parameters affect the objective and interdisciplinary variables that couple the subspaces. The goal of the system level optimizer is to select values that optimize the overall objective function. It also selects values for the interdisciplinary variables and recommends them as targets for each of the subspace optimizers to match within a set system level iteration. The subspace optimizer must ensure that its local disciplinary constraints are satisfied at each system level iteration [1].

There are three discipline level programs used in this study. The first discipline level is configuration and orbit design, and this module performs coverage analysis for different orbital parameters. To optimize this subspace efficiently, a combination of exhaustive grid search and heuristics were implemented [1].

The second disciplinary level program is spacecraft design, which estimates the mass and cost of the payload and spacecraft bus. The subspace optimizer is considered a hybrid optimization scheme, made of a univariate search and a gradient based search. This technique proved to have the ability to efficiently solve the spacecraft design optimization problem [1].

Launch manifest is the last disciplinary level program, and this module finds the minimum launch cost strategy to deploy the constellation. The subspace optimizer was created from a three step process. First, the integer programming problem was solved for the minimum cost strategy. Next, heuristics were used to select an option to improve the constraint values. Lastly, a univariate search completed the tradeoff between the mission orbit and spacecraft unit mass [1].

The need for a systematic, multivariable, multidisciplinary method was recognized and CO was applied as the approach for the system integration and optimization of a satellite constellation. The study included the design of the constellation's orbit, individual spacecraft, and deployment strategy. Successful

convergence related to the design and deployment problem was achieved, which

validated the use of CO for this type of problem [1].

2.3.2.4 STK-MATLAB Interface.

In this thesis, a constellation design tool is created using MATLAB and STK.

STK is a program that allows users to model, analyze, and visualize space systems.

Users can create objects such as satellites and constellations, as well as propagate specific

orbits through time. Once the orbits and constellations are defined, the program allows

users to quantify the performance using several different measurements. To create a

constellation design tool, MATLAB is used to execute all the commands needed to

operate STK in this scenario. The documentation in STK's programming help lists

specific STK operations in alphabetical order. This allows the user to search certain

commands. The documentation illustrates basic examples of the code, but without prior

knowledge of the format, it is difficult to debug the program.

2.3.3 Challenges in Designing a Constellation.

Designing a satellite constellation is a process, and Section 2.3.1 described the

steps. Since there are multiple factors included in the design of a constellation, several

challenges may exist. When determining the orbit type of the constellation, there are

advantages and disadvantages to each type. The challenge for the decision maker is

defining the most important factor and deciding if it is worth facing the limitations of

certain orbit types to accomplish the goal. Section 2.1.3 discussed the different

perturbations that exist with different orbit types, and these can drive the decision to use or not to use a certain orbit type. In constellation design, coverage is normally a key parameter because it is a fundamental element of performance [7]. However, according to Wertz, "Earth coverage is not a Gaussian parameter and statistical data can give very misleading results" [7]. This can make evaluating coverage a challenge, but there are methods used to overcome the challenge. Another challenge in constellation design is maintaining the required cost budget. With most space systems, a budget, along with requirements, is set before the design begins. Performance is an important factor, but as performance improves, the cost to support the system increases. Therefore, the designers must continuously be aware of the cost budget, so they do not design a constellation with a level of performance outside the customer's budget.

Within the cost budget, an estimate of launch cost is included. There are several challenges involved with estimating launch cost. The first challenge is that the cost of launch is flexible based on the law of supply and demand. In addition to the cost flexibility, there is potential for dual payload launches. For example, a larger launch vehicle may be selected, and the launch margin is used to accompany a secondary payload. This secondary payload can assist in the overall cost of the vehicle. Lastly, there are evolutionary trends in launch costs that dictate the price launch manufacturers can charge for certain vehicles. These challenges combined make it difficult to estimate the launch cost for a given system [7].

2.4 Optimization

Regardless of complexity, the general form of an optimization problem is to minimize a cost function that is subject to constraints. The solution is a vector of design variables that meets all the criteria [17]. In this thesis, there are two objective functions: PDOP and cost. A multi-objective genetic algorithm is selected as the optimization routine for this thesis. This section gives a description of numerical optimization, particle swarm optimization, and genetic algorithms.

2.4.1 Numerical Optimization.

When the number of variables and constraints in an optimization problem is greater than three, a numerical method is typically necessary to solve it. Numerical methods also have the ability to directly search for optimal points. There are several classifications of search methods used for nonlinear problems. First, derivative-based methods are used. They are based on the assumption that all functions of the problem are continuous and at least twice continuously differentiable [17]. Methods in this classification are also considered as gradient-based search methods. These methods use an iterative search process that is initiated with an estimate of the design variables. Gradients of the function are calculated using the values of the given function. The gradient is used to incrementally improve the solution, and it repeats until a stopping criterion is reached. These methods only use local information, so these methods always converge to a local minimum point, but they can be adjusted to find global solutions [17].

Direct search methods are another classification of search methods. These methods do not use derivatives of functions to find solutions. The only values that are considered are those of the functions. The methods are still able to work properly if the function values are unavailable; it just needs to be able to determine which point will lead to a better value compared to other points [17]. The next classification is the derivative-

2-28

free methods, which are methods that do not explicitly calculate the derivatives of functions. In order to create a local model, the derivatives are approximated using only the function values.

There are two types of optimization problems to consider: unconstrained and constrained. For unconstrained optimization, the basic iterative equation is shown in Equation 1.

$$x_i^{(k+1)} = x_i^{(k)} + \Delta x_i^{(k)}; i = 1 \ to \ n; k = 0,1,2, \dots \tag{1}$$

Where $x_i^{(k)}$ is the kth iteration of the ith design variable. The process is summarized as a general algorithm that consists of five steps. These steps are taken from [17]. First, estimate a reasonable starting design $x^{(0)}$ and set the iteration counter k=0. Second, compute a search direction $d^{(k)}$ at the point $x^{(k)}$ in the design space. This step needs a cost function value along with its gradient for unconstrained problems, and for constrained problems, this step needs the constraint function along with its gradients. Third, check for convergence of the algorithm; if it converged, then stop. Otherwise, continue the process. Fourth, calculate a positive step size α_k in the direction, $d^{(k)}$. Lastly, update the design as seen in Equation 2; set k=k+1. Then go back to the second step.

$$x^{(k+1)} = x^{(k)} + \alpha_k d^{(k)} \tag{2}$$

This general algorithm can be applied to unconstrained and constrained optimization problems, but with constrained problems, the constraints must be considered when determining the search direction. Arora gives a four step numerical algorithm for constrained problems [17]. First, linearize the cost and constraint functions about the current design point. Second, define the search direction using the linearized functions. Third, solve the sub-problem that gives a search direction in the design space. Lastly, calculate the step size to minimize a descent function in the search direction [17].

The algorithms described above are basic algorithms that can be used in numerical optimization. It is important to understand the type of problem being solved and ensure that the chosen algorithm is well-suited for the problem. Without the appropriate algorithm, it is not guaranteed that accurate solutions will be generated.

2.4.2 Particle Swarm Optimization.

Particle swarm optimization (PSO) is another method that can be used for constellation design. There are some common terms used in this optimization process. The population of the algorithm is considered a swarm, and a particle is an individual member of the swarm that represents a possible solution to the problem. A leader is a particle that is used to guide other particles towards more efficient regions of the search space. The velocity vector is the vector that drives the optimization process by defining the direction for a particle to travel in order to improve its current position. To control the impact the previous history of velocities has on the current velocity of a particle, inertia weight is used. The learning factor represents the attraction that a particle has toward either its own success or that of its neighbors [18].

Having an understanding of the terms used in this process aids in the comprehension of the PSO algorithm. The PSO algorithm uses fewer parameters compared to GAs, and it does not use the operators such as crossover and mutation. It is easier to execute on computers because it does not require the use of binary number encoding or decoding. According to Reyes-Sierra *et al.*, GAs utilize three mechanisms (Section 2.4.3), and the PSO only uses two. There is no specific selection function and no offspring generation as in the GA. PSO uses leaders with specific velocities and directions to guide the search, but the GA uses fitness values and offspring to improve individuals. Figure 2-7 illustrates the overall algorithm for PSO [19].

```
Begin
    Initialize swarm
    Locate leader
    g = 0
    While g < gmax
            For each particle
                    Update Position (Flight)
                    Evaluation
                    Update pbest
            EndFor
            Update leader
            g++
    EndWhile
End
```

Figure 2-7: PSO Algorithm

In [20] the performance of the PSO with an adaptive inertia weight was studied, and the results of several test cases were illustrated. This research utilized four nonlinear functions as their test cases: the Sphere function, the Rosenbrock function, the Rastrigrin function, and the Griewank function. The Sphere function is described by Equation 3:

$$f_0(x) = \sum_{i=1}^{n} x_i^2 \qquad (3)$$

Where $x = [x_1, x_2, ..., x_n]$ is an n-dimensional real-valued vector. The second function, the Rosenbrock function is shown in Equation 4:

$$f(x)_1 = \sum_{i=1}^{n}(100(x_{i+1} - x_i^2)^2 + (x_i - 1)^2) \qquad (4)$$

The third function is the generalized Rastrigrin function described by Equation 5.

$$f_2(x) = \sum_{i=1}^{n}(x_i^2 - 10\cos(2\pi x_i) + 10) \qquad (5)$$

The last function, described in Equation 6, is the generalized Griewank function.

$$f_3(x) = \frac{1}{4000}\sum_{i=1}^{n} x_i^2 - \prod_{i=1}^{n}\cos\left(\frac{x_i}{\sqrt{i}}\right) + 1 \qquad (6)$$

For the population initialization, they used what was considered an asymmetric initialization method. In all four test cases, the shapes of the curves from the data show that the PSO converged quickly. However, it began to slow its convergence speed down as it approached the optima. This was attributed to the use of the linearly decreasing inertia weight, which prevented the PSO from having global search ability at the end of the case. It was concluded that by using the linearly decreasing inertia weight, the performance of the PSO could be improved greatly when compared to the evolutionary algorithm used in [21]. Other conclusions made from this research were that the PSO resulted in similar performance with different population sizes and it scaled well for all four functions [20]. This research demonstrates the successful use of the PSO on several different nonlinear functions.

2.4.3 Genetic Algorithms.

An understanding of genetic algorithms (GA) is imperative to this thesis because it is the optimization method used in this design tool. GAs are based on the genetic processes of biological organisms. The basic structure of a GA consists of a randomly selected population of individuals. These individuals represent a point in the design space of possible solutions. According to Arora, "The basic idea of a GA is to generate a new set of designs (population) from the current set such that the average fitness of the population is improved" [17].

There are several different terms that make up the GA process. A population is a set of design points at the current iteration. Once the population is defined, the calculations are completed to create a new generation. The tolerance is defined as the smallest change in the objective function between generations. A chromosome is a synonym in a GA for an individual design point. Lastly, there is a gene, which is a scalar valued component of the design vector [19].

There are three main parts that make up the GA: the objective function, the variables, and the constraints [22]. However, in this research, there are multiple objective functions, and the MATLAB multi-objective function GA cannot process non-linear constraints. Therefore, in this thesis, the objective functions and design variables (gene) are the only parts that make up the GA.

To implement a GA, an initial population is required. This population creates a design space for the algorithm to work in. After each individual in the population is evaluated by the value of the fitness function, a selection process is used to select a group of parents. The parents are selected from the individuals with the lowest (this research is a minimization problem, so lower values are better) fitness values in the population [22]. The parents are then paired off, and the string content is swapped between the parents in a pair during the crossover process [17]. Figure 2-8 illustrates the process of cutting chromosomes at a determined point and swapping the tails (crossover). The resulting chromosomes from this process are called children [23].

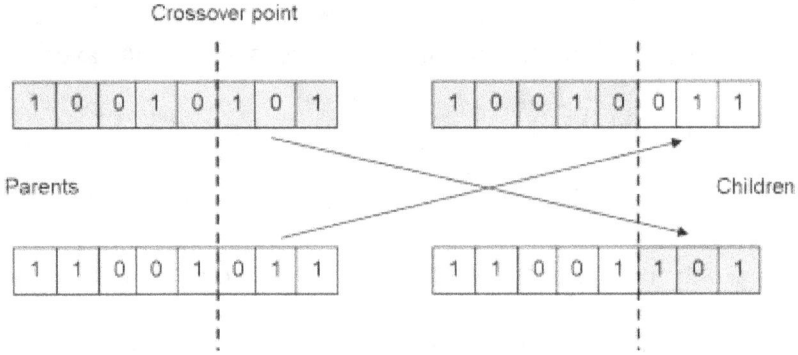

Figure 2-8: Crossover Operator

After the crossover operation, the algorithm will perform a mutation operation. The mutation operation is a probabilistic modification to the encoding of an individual chromosome [17]. Figure 2-9 demonstrates the process of flipping the value of a bit at a determined point. The resulting chromosomes from this operation are called children as well [23].

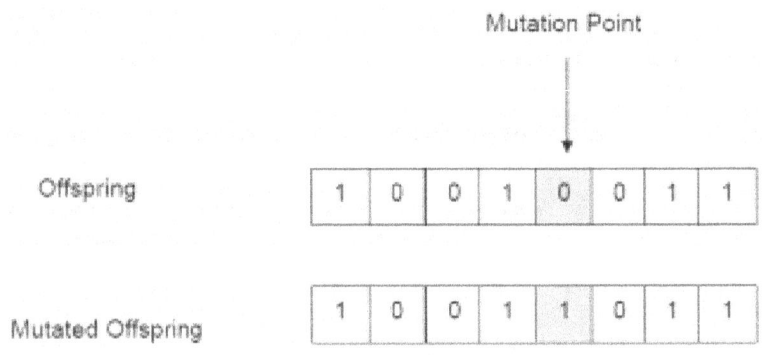

Figure 2-9: Mutation Operator

After the crossover and mutation operations are completed, a new generation is produced. The new generation consists of a higher proportion of the genes possessed by the elite members of the previous generation [17]. Through this process, the strong characteristics are spread throughout the population over several generations. By mating

2-34

most fit individuals in a population, the regions of the design space with the best chance

for an optimal solution are explored. Figure 2-10 illustrates the overall process for a GA

[23].

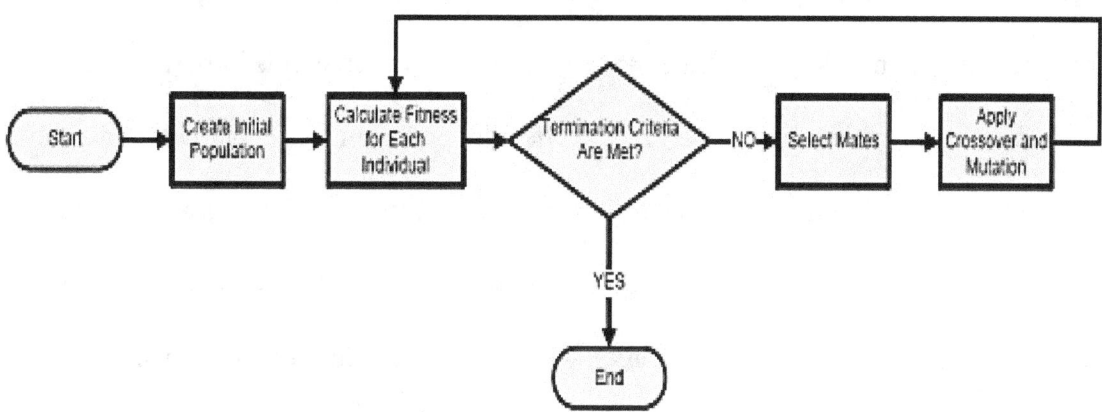

Figure 2-10: GA Flowchart

2.4.4 Computational Concerns.

GA's can be applied to a wide range of problems, and they are fairly simple to use

because they do not require the use of gradients. However, there are several drawbacks

in using this algorithm. Even with a reasonably sized problem, GAs require a large

amount of computation time, especially with problems that require massive calculations

for the function values. In this thesis, STK is performing the PDOP calculations. As the

number of satellites in the scenario increases, the time it takes for STK to compute all the

accesses increases as well. Since the time to calculate the function value increases, it

adds to the overall computation time of the GA. The GA does not guarantee a global

solution, but it can be overcome by running the algorithm several times or by allowing it

to run for longer periods. This adds to the computational concern because with larger generations and population sizes, the algorithm takes increasingly more time to compute solutions [17].

The computation time does pose concern for problems using this algorithm because without the proper amount of time, it is not guaranteed that accurate results will be determined. In order to mitigate the large computation time, it is important to understand the problem being analyzed and develop a design space that will allow the GA to function efficiently. For this research, several different test cases are used to analyze the design tool, so having multiple computer systems to operate the algorithm alleviates some of the concern associated with the computation requirements.

2.4.5 Multi-Objective Genetic Algorithms

Multi-objective genetic algorithms are built on the same foundation as the GAs, so the procedures discussed in Section 2.4.3 still apply. There are two general approaches to multi-objective optimization problems: combine the single objective functions into a combined function or determine a Pareto optimal set [24]. A Pareto optimal set is a set of solutions that cannot improve any of the objective function values without sacrificing the others. The overall goal of a MOGA is to produce a Pareto optimal set, but identifying the entire set is infeasible due to its size. Therefore, the practical method is to analyze the best-known Pareto set.

The concept of investigating a Pareto optimal set is often referred to as Pareto optimality. Arora defines this as, "A point x^* in the feasible design space S is Pareto optimal if and only if there does not exist another point x in the set, S such that $f(x) \leq f(x^*)$ with at least one $f_i(x) < f_i(x^*)$." Figure 2-11 illustrates dominate and non-dominate solutions along with the Pareto front. Pareto optimal solutions are also referred

to as non-dominated solutions, which means no other solution is better than them in all

the objectives [25]. In this thesis, the objective functions are cost and PDOP, and it is

desirable to minimize both of these functions.

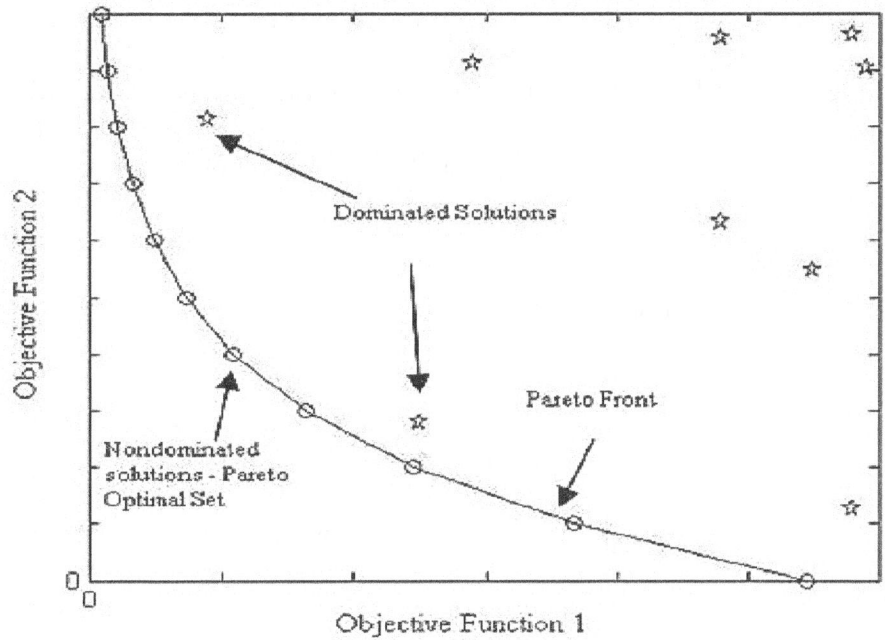

Figure 2-11: Pareto Front

With Pareto optimality, a solution with high cost and low PDOP values can be

Pareto optimal, along with a solution with low cost and high PDOP values [2]. To select

designs for later generations, the MOGA possesses different selection techniques. The

method used in this work is considered ranking. This process takes a given set of designs

and evaluates the objective functions at each point. Then each point is determined to be

either dominated or non-dominated based on the comparison of the vector of objective

function values at the given point [17]. Non-dominated points are given a rank of one

and then removed from consideration. The points that are non-dominating relative to the

remaining group receive a rank of two. This process continues until all the points are

2-37

ranked. The points with the lowest rank possess the highest fitness value, and the points with high fitness values are selected for the next generation [17].

Since many real world engineering problems require the ability to analyze tradeoffs between multiple functions, the MOGA is a beneficial tool for optimization. The computation requirements do increase with multiple objective functions, so the user must be aware of the realities of the algorithm. Also, customization of the MOGA approach is necessary in order to properly handle the different objective functions being analyzed. It is not required to determine every Pareto optimal solution. However, it is essential to identify Pareto optimal solutions across a certain range of interest for the objective functions [24].

2.4.5.1 GA Based Constellation Design.

An example of the use of a GA for constellation design is illustrated in [25]. The design of optimal satellite constellations is addressed. A genetic algorithm termed, Modified Illinois Non-dominated Sorting Genetic Algorithm (MINSGA) was applied to several test cases to show the Pareto optimal region for multiple objectives. The MINSGA was combined with STK to produce several constellation designs that yielded continuous global coverage.

The Pareto GA is based on the concept of non-dominating sorting. The MINSGA replaces the stochastic remainder with stochastic universal selection. The MINSGA was verified through several test cases and compared to the Non-dominated Sorting Genetic Algorithm (NSGA) from [26]. Six different test cases were used to demonstrate the accuracy and ability of the MINSGA to determine the Pareto optimal region. The last three test cases were used to demonstrate the advanced capabilities of the algorithm.

STK was used to calculate the values of the objective functions for each individual in the GA population. The combination of the MINSGA and STK was given the term CODEC meaning Constellation Optimal Design by Evolutionary Computation. Five different test cases were used to establish the performance of CODEC. The first test case is called Inclination Characterization, which determines the inclination range for a Walker 5/5/1 configuration. CODEC correctly identified the region between 15 and 90 degrees inclination as providing global coverage. The second test case is called Walker-Delta Configuration Study. This case included the Walker configuration as part of the parameter set. The goal was to demonstrate CODEC's ability to handle multi-objective optimization because this case analyzed the trade between number of satellites and percent coverage. Three solutions resulted from the test case, and all three provided global coverage.

The third test case is called Open Structure with Fixed Inclination, and this case addressed continuous global coverage versus number of satellites but without restricting the constellation to a Walker Delta pattern. The solution provided 99% continuous global coverage with a small portion of the poles out of view for a short period each day. The fourth test case is Open Structure with Posigrade Inclination, which involved the same scenario from the third case. However, the inclination was allowed to assume a posigrade value. The solution from this analysis provides 99.7% continuous global coverage, with short coverage gaps near the poles, and with more time to evolve, the solution was expected to reach 100% coverage. Lastly, the fifth test case is called Open Structure with Open Inclination, and this test case allowed the inclination to vary between 0 and 180 degrees. This configuration provides 99.2% global coverage, but it is not a desirable configuration because the satellites possess retrograde inclinations. Although the configuration is not desirable, it demonstrated the ability of the algorithm to create novel solutions to the problem [25].

The use of a Pareto genetic algorithm to perform satellite constellation design was demonstrated. It showed that a group of solutions can be created, and it illustrated tradeoffs between performance and cost. It also illustrated the success of incorporating pre-existing commercial software (STK) with a genetic algorithm.

In [27] a genetic algorithm was applied to a different problem. A genetic algorithm was used to generate sets of constellation designs. Pareto fronts were used to show tradeoffs for two conflicting functions.

A baseline constellation design from William *et al.* was chosen and reconstructed using a multi-objective genetic algorithm [28]. The objective functions for the baseline research were to minimize both the maximum revisit time and the area-weighted average revisit time to a region of points representative of the entire Earth. After the sparse-coverage tradeoff analysis, a resolution tradeoff was constructed. The temporal verses spatial resolution tradeoff for constellations of Earth-observing satellites was analyzed.

The baseline constellation was used for comparison, and the GA was used to determine the tradeoffs. The GA approximated Pareto fronts for several different constellation designs. The results matched those of William *et al.* in most cases, but for several designs, the GA was able to design constellations that performed better (lower maximum and area-weighted average revisit time) [28]. The fronts calculated possessed, on average, a greater spread and consisted of more non-dominated designs. Since the replication of the baseline problem was successful, the tradeoffs between spatial and temporal resolution were analyzed.

A multi-objective conflict arises as the altitude of a LEO satellite increases; the image quality decreases while the revisit time is improved. The motivation in this study was to discover the sensitivity of the constellation design's maximum revisit time to a small change in altitude. With the test cases, it was discovered that a small change in altitude can affect the maximum revisit time negatively. However, as the number of satellites increases, the discontinuities decrease, and the penalty paid in maximum revisit time is less severe.

Important considerations when applying a multi-objective genetic algorithm to a constellation design problem were highlighted. The several approximated Pareto fronts that were generated were able to find designs that improved the objective functions significantly. The random seed had the greatest impact on the results of the final generation, and to improve the number of generations needed, the multi-objective genetic algorithm must be less sensitive to the initial seed value. It was also found that maximum revisit time was highly sensitive to small changes in altitude, and the resolution fronts exhibited discontinuous, nonlinear characteristics. Overall, the use of a multi-objective genetic algorithm as a constellation design tool and the benefits of using a Pareto front to analyze tradeoffs in the objective functions were illustrated [27].

2.5 Navigation Metrics

One of the most important concerns with the GPS constellation is the accuracy of its positioning and timing measurements. The accuracy of a navigation constellation is the true measure of its capability. Position accuracy is determined by the geometry of the satellites overhead and accurate pseudorange[7] measurements [4]. Along with analyzing the geometric errors of the constellation, signal strength is another factor that is necessary in the development of a navigation constellation. Having the appropriate amount of transmit power is essential to ensuring suitable signal strength at the receiver. The following sections will describe these topics.

2.5.1 Geometric Errors.

The satellite geometry in relation to the receiver at the time the signal is received affects the positon solution. Dilution of Precision (DOP) is the term that describes the geometry of the satellites. When satellites are efficiently distributed in the sky, the geometry is good for determining position, but if the satellites are too close together it results in poor geometry over the receiver [29]. With good geometry, the DOP values are low, and for poor geometry, the DOP values are high. Figure 2-12 illustrates the concept of good and bad satellite geometry [29].

---

[7] Pseudorange is simply the observed signal propagation delay scaled by the speed of light in a vacuum.

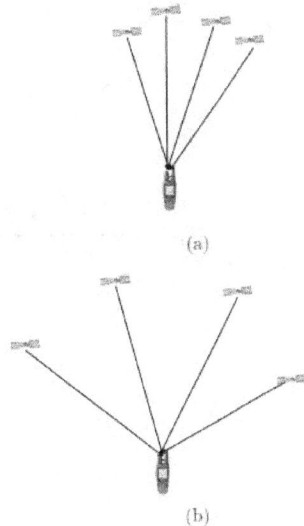

Figure 2-12: a) Poor satellite geometry b) Good satellite geometry

In order to determine receiver position in three dimensions (x_u, y_u, z_u) and the

clock offset t_u, pseudorange measurements should be made to four satellites which can be

written as follows:

$$\hat{\rho}_k = \sqrt{(x_k - \hat{x}_u) + (y_k - \hat{y}_u) + (z_k - \hat{z}_u)} + ct_u \qquad (7)$$

Where $\hat{\rho}_k$ is the approximate pseudorange between the receiver and the kth satellite, x_k is

the kth satellite position vector, and $\hat{x}_u$ is the approximate receiver position vector. These

nonlinear equations can be solved through iterative techniques based on linearization. If

an estimate of the receiver position is known, the position offset is denoted

as Δx_u, Δy_u, Δz_u. The receiver position consists of the approximate position plus the

position offset. The linearized form of Equation 7 is shown below [23].

$$\Delta \rho_k = a_{xk}\Delta x_u + a_{yk}\Delta y_u + a_{zk}\Delta z_u - c\Delta t_u \qquad (8)$$

Where,

$$\Delta\rho_k = \hat{\rho}_k - \rho_k$$

$$a_{xk} = \frac{(x_k - \hat{x}_u)}{\hat{r}_k}$$

$$a_{yk} = \frac{(y_k - \hat{y}_u)}{\hat{r}_k}$$

$$a_{zk} = \frac{(z_k - \hat{z}_u)}{\hat{r}_k}$$

$$\hat{r}_k = \sqrt{(x_k - \hat{x}_u)^2 + (y_k - \hat{y}_u)^2 + (z_k - \hat{z}_u)^2}$$

These equations can be expressed in matrix form [23]:

$$\Delta\rho = \begin{bmatrix} \Delta\rho_1 \\ \Delta\rho_2 \\ \vdots \\ \Delta\rho_n \end{bmatrix} \qquad H = \begin{bmatrix} a_{x1} & a_{y1} & a_{z1} & -1 \\ a_{x2} & a_{y2} & a_{z2} & -1 \\ \vdots & \vdots & \vdots & \vdots \\ a_{xn} & a_{yn} & a_{zn} & -1 \end{bmatrix} \qquad \Delta x = \begin{bmatrix} \Delta x_u \\ \Delta y_u \\ \Delta z_u \\ c\Delta t_u \end{bmatrix}$$

Using the least-squares solution, an estimate for x is determined:

$$\Delta x = (H^T H)^{-1} H^T \Delta\rho \qquad\qquad (9)$$

The $(H^T H)^{-1}$ matrix is the DOP matrix, and it relates the measurement errors to the position errors. The elements of $(H^T H)^{-1}$ are designated as:

$$(H^T H)^{-1} = \begin{bmatrix} \sigma_x^2 & \sigma_{xy} & \sigma_{xz} & \sigma_{xt} \\ \sigma_{xy} & \sigma_y^2 & \sigma_{yz} & \sigma_{yt} \\ \sigma_{xz} & \sigma_{yz} & \sigma_z^2 & \sigma_{zt} \\ \sigma_{xt} & \sigma_{yt} & \sigma_{zt} & \sigma_t^2 \end{bmatrix} \qquad\qquad (10)$$

It is possible to examine specific components instead of the overall solutions. The specific components include: the three-dimensional receiver position coordinates, the horizontal coordinates, the vertical coordinates, or the clock offset [30]. These components can be summarized by five DOP categories which include geometric (GDOP), horizontal (HDOP), vertical (VDOP), position (PDOP), and time (TDOP). In this thesis, PDOP is the DOP category that is used to measure performance. It is represented in Equation 11.

$$PDOP = \frac{1}{\sigma}\sqrt{\sigma_x^2 + \sigma_y^2 + \sigma_z^2} \tag{11}$$

When multiplied with the root mean square value, PDOP represents the magnitude of the total position error in all three dimensions.

2.5.2 Signal Strength.

Even though a navigation constellation is able to provide strong geometry over its receivers, it is important to ensure the signal is strong enough to be interpreted by its receivers. Each receiver possesses a certain level of sensitivity, so the signal power must be above that level in order for the system to be decoded properly. In this thesis, transmit power is a design variable that is used to determine satellite payload mass. It is set at given altitudes to ensure the link is closed with each of the constellation designs. Link closure is described as "the positive allocations (power) must balance with negative allocations (attenuation and other losses)" [7].

To complete a link analysis, the budget entries that account for signal power must be analyzed along with noise factors. The first section is the transmit power along with transmitter output losses. The main product of this section is the equivalent isotropic radiated power (EIRP), which equals the power of the transmitter plus the gain of the transmitter minus losses related to the hardware between the transmitter and antenna.

The second segment is the free-space and atmospheric losses (L_s), which is the largest

loss factor. It depends on the signal wavelength and satellite altitude, so the higher

altitude, the larger the losses.

The third section is the received power, which is dependent on the EIRP, gain of

the receiver, and associated losses from the atmosphere and hardware. Next, all

significant sources of noise must be accounted for, and it is commonly referred to as

system noise temperature (T_s). Fourth, the ratio of the receiver gain and T_s is determined,

and this factor is used to calculate the signal to noise ratio. The next section determines

the carrier-to-noise ratio, which is the ratio of the received signal power to noise. The

final step in the link analysis is to divide the carrier-to-noise ratio by the data rate to

obtain $\frac{E_b}{N_o}$. Each receiver has a required $\frac{E_b}{N_o}$ value, so the link margin is simply the

predicted value minus the required value [7]. When the value is positive, it is considered

positive link margin, and if it is negative, the signal strength is too low for the system to

function properly. The link is considered closed when the margin is zero. For this

research, the required receiver signal strength was known for GPS, so this process was

rearranged to determine the transmit power needed at various altitudes in order to close

the link. See Appendix A-1 for details on the equations.

2.6 Previous and Current Work

Satellite constellation design is a vital process when it comes to the success of

specific missions. There are multiple types of constellations, all of which are applied to

different missions. The success of other projects in this field allows the design tools that

already exist to be improved and to be applied to new scenarios.

2.6.1 Constellation Design.

At the Air Force Institute of Technology (AFIT), Major Robert Thompson applied a multi-function, multi-orbit disaggregated space system optimization methodology to the space-based defense weather enterprise [31]. A GA was used as the optimization technique, and he utilized this method to assess and compare alternate space-based weather system conceptual architectures. The overall objective of his analysis was to illustrate the applicability of the developed Disaggregated Integral Concept Optimization (DISCO) methodology to assess multi-function, multi-orbit disaggregation problems.

The individuals used in his research consisted of the variables (genes): satellite planes, satellites per plane, satellite orbital height, sensor aperture diameter, sensor view angle, maximum vertical cell size, and number of launch vehicles. The optimization model for the Weather System Follow-on (WSF) is attempting to minimize cost while subject to performance constraints. The assessment models are subject to unmanned large and small satellite cost assessment models, dynamics coverage models, and sensor performance models. A GA global optimization routine integrates the optimization and assessment models. The results of his research demonstrated the applicability of the DISCO methodology to general multi-function disaggregated problems [31].

Following on to the research conducted by Major Thompson, Second Lieutenant (2d Lt) Evelyn Abbate, graduate of AFIT, also performed research using a GA [19]. The optimization model used in her research consisted of minimizing cost of a disaggregated constellation while subjecting the model to certain performance constraints. She used Walker parameters, orbital elements, and sensor size as her design variables (genes). Each of these variables was applied to both large and small satellites. The cost model in her system was the same as the cost model used in this research, but the payload mass calculation is different in this thesis.

The results of her work demonstrated the successful use of a design tool consisting of MATLAB's single objective GA and STK to optimize the cost of disaggregated constellations over a given target deck. The results also illustrated that a constellation of large satellites, augmented with smaller satellites, could increase performance capabilities. A constellation made up of small satellites could provide consistent coverage at a minimum National Image Interpretability Rating Scale (NIIRS) level of three with just a few large satellites to augment it.

2.6.2 Navigation Systems.

Although previous research has been performed in the area of constellation design, other research has been done in the area of navigation systems. Major Bryan Bell, an AFIT graduate, focused specifically on the GPS constellation and the performance of the constellation in a degraded state [32]. He utilized STK to evaluate the PDOP values of alternate constellations and the combination of constellations to augment a degraded GPS state. The main objective for his work was to develop a foundation for a basic Implementation Plan for restoring navigation capability.

Major Bell concluded from his research that the nominal GPS constellation of 24 satellites can maintain an average PDOP less than six when two orbital planes are degraded. However, if three or more orbital planes are degraded, the average PDOP is greater than six, which is not acceptable for mission success. When augmented with the 66-satellite Iridium constellation, the average PDOP remains less than six for losses of four or fewer planes. Average PDOP values are only greater than six when more than

four planes are degraded when GPS is augmented with a highly elliptical orbit (HEO) 4/1

Walker. Lastly, the best PDOP values resulted when GPS was augmented with a GEO

constellation of three satellites; there were no cases when four or fewer planes were lost

where the average PDOP exceeded six. His research demonstrated the improvement with

average PDOP values when the GPS constellation was augmented with other

constellations [32]. The tool presented in this thesis analyzes navigation constellation

designs, and if combined with the results of Major Bell's research, it could determine

possible augmentation constellations for GPS.

2.7 Applications

At AFIT, the Spacecraft Systems Engineering course required several student

groups to develop reports on their specific topic. The Global Navigation Satellite System

(GNSS) Design Group developed a project called G^3, which represents GPS, GLONASS,

and Galileo. This group was tasked to augment the current GPS with a low cost satellite

that is capable of providing a variety of signals in LEO [33]. The tool designed and

analyzed in this thesis can be applied to the problem addressed in the G^3 report. The G^3

project performed an orbit selection analysis to determine the best orbit that met the

requirements given in the problem. The design tool in this thesis can be used to select an

appropriate orbit design that minimizes cost and PDOP for a given altitude. The group

had to place constraints on their analysis to meet the requirements. The design tool in

this thesis cannot apply constraints, but it would aid in the analysis of what altitude is

optimal and the resulting PDOP from the design. From that point, the group would be able to analyze the design further to ensure the appropriate requirements are met for the system. This design tool could also aid in developing their project further by analyzing the tradeoffs of augmenting the current GPS with a constellation of satellites in LEO. It would provide a method for determining the most efficient Walker and orbital parameters for the design.

2.8 Summary

This thesis will use a MOGA to generate navigation constellation designs and analyze the tradeoffs between PDOP and cost. Having knowledge of astrodynamics and constellation types will help in understanding how the design variables (genes) affect the constellation designs. Discussing the optimization methods and examples illustrated other studies related to constellation design. The geometric errors and signal strength description demonstrated the background for the PDOP calculation and the method in which to determine the signal strength for a system. The next chapter will apply these concepts by detailing the specific models used in this thesis.

Chapter 3
Methodology

The Air Force is looking for new methods to provide navigation that are less

costly but still provide accurate positioning and timing. The design tool presented in this

thesis provides optimal batch Pareto fronts and illustrates the tradeoffs of the objective

functions. This chapter outlines the methodology used to develop the constellation batch

Pareto fronts. Validation of the design tool is demonstrated using the current GPS

constellation.

3.1 Problem Statement

This research uses MATLAB's MOGA along with STK to explore the trade space

for navigation constellation designs at different orbital altitudes. The objective is to

analyze the tradeoffs between PDOP and cost of the different solutions. Different design

guidelines are inferred from using the optimized constellations. Different test cases are

analyzed to determine an optimal navigation constellation design at LEO, MEO, and

GEO. Using those results, hybrid navigation constellation designs are explored.

3.1.1 Design Variables

The MOGA determines the design vectors that reside within the specified bounds

of each scenario. The design variables for this research are separated into three

categories: Walker parameters, orbital parameters, and transmit power. The bounds

varied with different altitudes to ensure STK had enough satellites to calculate PDOP. If

the lower bounds of planes and satellites per plane are too low, STK is unable to

determine PDOP, and the tool does not run properly. Determining the appropriate

3-51

bounds for different test cases requires input from the designer to analyze the solutions and adjust the bounds to produce more accurate results. Table 3-1 summarizes the bounds of the parameters for the specific test cases used in this research.

Parameters	Units	Bounds-Validation	Bounds LEO Case1	Bounds LEO Case2	Bounds LEO Case3	Bounds MEO Case1	Bounds MEO Case2	Bounds MEO Case3	Bounds MEO Case4	Bounds GEO Case
#Planes	-	5-8	10-15	9-15	8-15	6-15	3-11	4-11	3-9	2-10
Sats/Plane	-	2-6	10-15	9-15	8-15	6-15	3-11	4-11	3-9	3-10
Truan	deg	5-10	0-180	0-180	0-180	0-180	0-180	0-180	0-180	0-180
RAAN Inc	deg	50-70	0-180	0-180	0-180	0-180	0-180	0-180	0-180	0-180
Alt	km	20180-20280	725-825	1150-1250	1575-1675	2000-2100	10447-10547	18893-18993	27340-27440	35786-35796
Incl	deg	40-60	40-80	40-80	40-80	40-80	40-80	40-80	40-80	30-80
TX Power	W	48.5	0.5	0.97	1.28	1.83	16.12	30.07	94.83	128.99

Table 3-1: Design Parameters

Transmit power is included as a design variable that is directly related to payload mass. With more power, the payload mass, spacecraft mass, and system cost increase. Because transmit power is directly related to cost, a method is developed to ensure transmit power also is related to PDOP. Without developing a constraint for transmit power, the design tool produces solutions at high altitudes with small satellites because it results in lower costs. To mitigate this problem, several test cases were created based on different altitudes. Using the ranges of altitudes given in Section 2.1.1, a quarter of the distance from the beginning of LEO to the beginning of MEO was calculated as 425 km. This was used as the step size within the LEO range. The test cases were built starting at 300 km and increasing by the calculated step size. However, the case at 300 km is not included due to computational limitations. The lower altitude possessed many satellites which increased the computational burden for the PDOP calculations. This caused errors that prevented the MOGA from computing. Therefore, the first test case is at an altitude of 725 km.

The same process was used to determine the step size within the MEO range, and it was calculated as 8447 km. Once the altitudes for the test cases were determined, the calculations for link margin, described in Section 2.5.2, were used to determine the amount of transmit power needed at a given altitude to close link at the edges of the Earth. This method limits the MOGA from varying transmit power and altitude, but it constrains the size of the satellites for the appropriate altitudes and guarantees link closure. Since altitude is a large factor in both the cost and PDOP functions, the MOGA would only produce a couple design solutions on the Pareto front when altitude was set to a single value. Therefore, to produce an accurate Pareto front with good distribution in points, a small range in altitude is used for each test case, but the appropriate transmit power is also included.

3.1.2 Assumptions

The batch Pareto fronts produced in this research depend on several assumptions. The calculation of PDOP, which is performed in STK, is an adequate metric for the scenario. When STK calculates PDOP, it only offers users a couple options for the values: minimum, maximum, average, or percent below. To capture the worst case values, the maximum values are determined for each global grid point over the simulation interval. This research is analyzing the global PDOP value, so a single value for PDOP is needed. Therefore, the median of the maximum values is used as the PDOP metric. The median is used to measure the central tendency because the data distribution at lower altitudes is positively skewed. Figure 3-1 illustrates the probability density function of PDOP in LEO. The median give a more accurate measure of the central tendency for that data. The mean resides towards the lower end of the distribution, which is not the true measure of the central tendency. Figure 3-2 and Figure 3-3 illustrate the probability density function for PDOP in MEO and GEO, respectively. These figures demonstrate the median and mean values are either the same or very similar. Therefore, for those cases, either measure would demonstrate the central tendency of the data. The median was chosen as the measure because of its ability to show the central tendency for each of the three distributions.

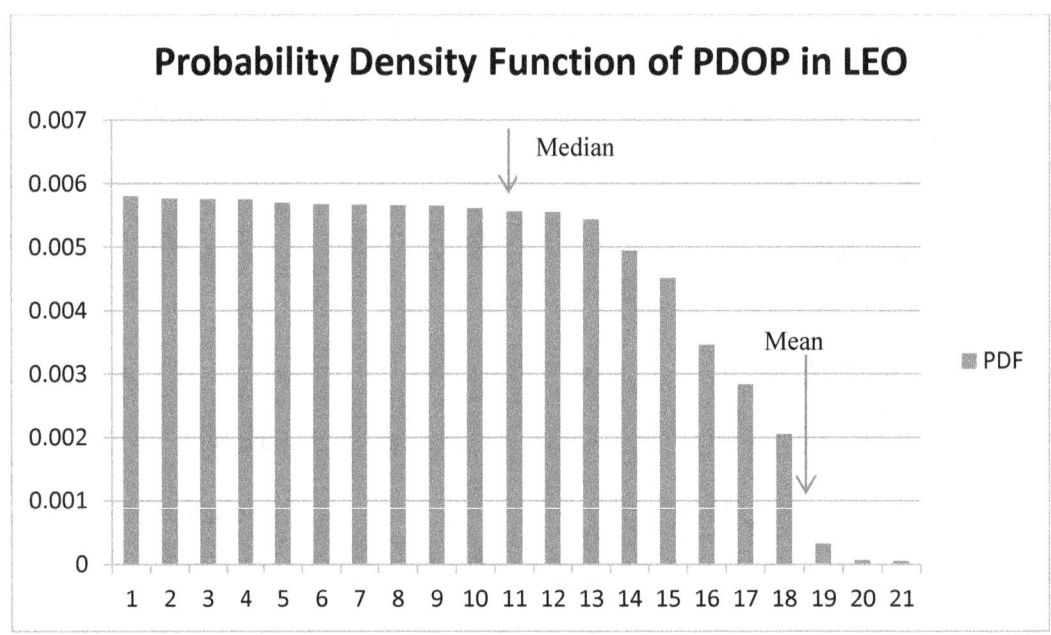

Figure 3-1: Probability Density Function of PDOP in LEO

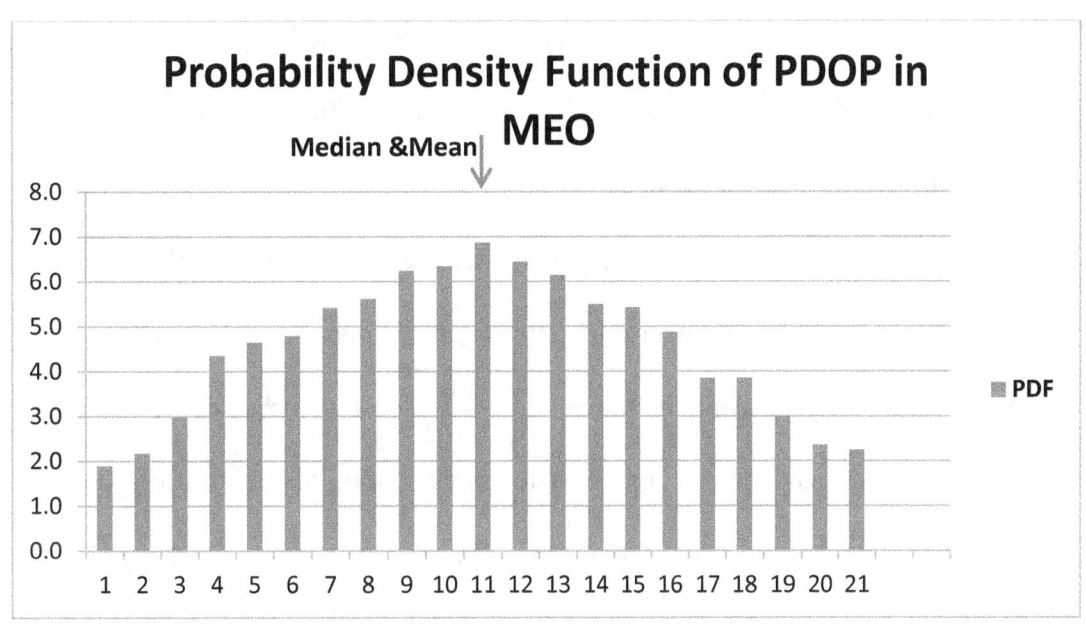

Figure 3-2: Probability Density Function of PDOP in MEO

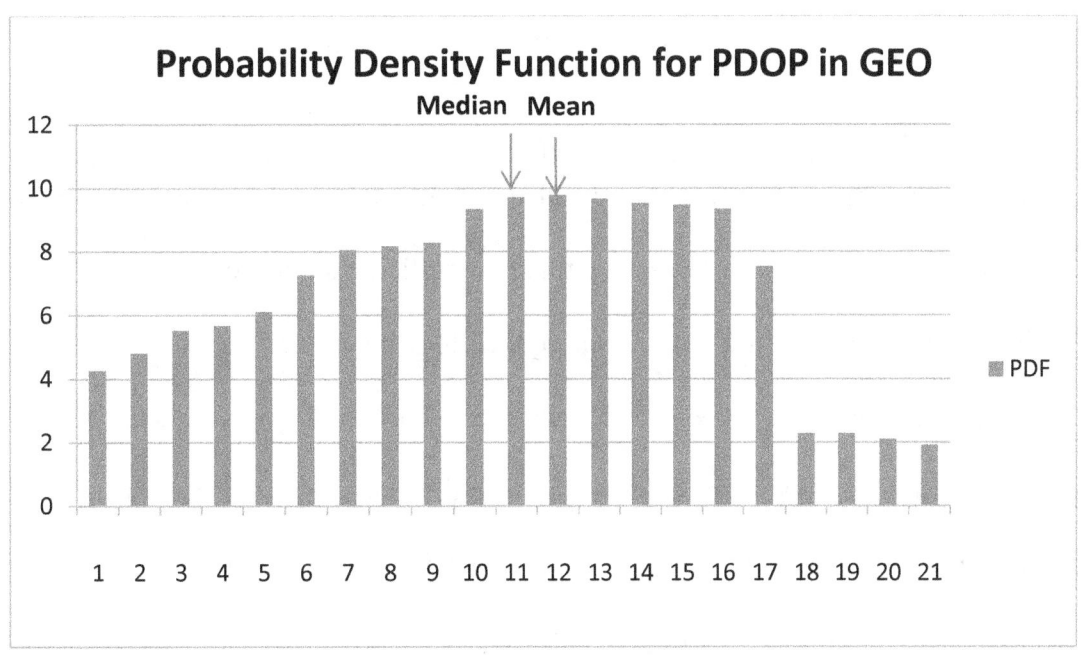

Figure 3-3: Probability Density Function of PDOP in GEO

It is also assumed in the PDOP calculation that the coverage grid is sufficiently dense. There must be a balance between resolution and time. Better resolution allows for more accurate results, but the computation time is increased. The PDOP calculation assumed no terrain limits, which is not necessarily realistic but cannot be directly addressed. Real world includes skyscrapers and mountains that affect the satellites that are visible to a user on the ground.

For the cost model, the costs are in fiscal year 2010 dollars (FY2010$). All models are imperfect, and the cost models are approximations that should only be compared to solutions using the same models. It is also assumed in the cost model that a communications payload is an appropriate representation of a navigation payload. When determining launch vehicle cost, the average launch costs from [7] are used. However, the costs are separated into LEO and GEO. Therefore, it is assumed that any satellite at an altitude higher than 2000 km will use the launch vehicles available at GEO. The only available launch vehicles are Pegasus XL, Minotaur IV, Falcon 9, Atlas 5, and Delta 4 Heavy.

All orbits that are developed and analyzed in this research are forced to be circular. It is assumed that this will still show sufficient trends to develop guidelines. When calculating the transmit power required for each of the test cases, it is assumed that each satellite is equipped with a GPS antenna. This assumption allows the use of the power and gain values from a GPS satellite in the link margin calculations. Table 3-2 summarizes the specifications for GPS at three different elevation angles. The values for five degrees elevation were used in this analysis. The shaded rows were not used in the calculations because several of the values were not needed in the link analysis. Others, such as the range and path loss, were calculated based on altitude. The only losses considered in the link budget are those mentioned in Section 2.5.2. For each test case, a required transmit power is determined, but each test case needs a small range in altitude in order for the MOGA to function properly.

	Satellite at 5° Elevation	Satellite at 40° Elevation	Satellite at 90° Elevation
Range(km)	25240	22020	21190
Satellite Antenna Gain, dB	12.1	12.9	10.2
Effective Isotropic Radiated Power, dBW	26.4	27.2	24.5
Path Loss, dB	-159	-157.8	-157.1
Atmospheric Loss, dB	0.5	0.5	0.5
Received Power Density, dBW/m^2	-133.1	-131.1	-133.1
Effective Area of an Omnidirectional Antenna, dBm^2	25.4	25.4	25.4
Receive Power Available from an Isotropic Antenna, dBW	-158.5	-156.5	-158.5
Gain of a Typical Patch Receive Antenna, dBic	-4	2	4
C/A Code Received Power Available to a Typical Receive Antenna, dBW	-162.5	-154.5	-154.5

Table 3-2: Gain and Power Specifications for GPS [34]

3.1.3 Objective Functions

For this research, PDOP and cost are the objective functions. To ensure that each design results in PDOP values that can produce accurate positioning and timing, constellation designs with PDOP values less than six are analyzed. The threshold of six is used to distinguish between desirable and poor constellation geometries.

The cost model used in this thesis consists of the Unmanned Space Vehicle Cost Model, version 8 (USCM8) Theoretical First Unit (TFU), Non-Recurring Engineering (NRE), and Small Spacecraft Cost Model (SSCM) [7]. The USCM8 derives cost estimating relationships (CER) from a total of 44 satellites using statistical regression techniques to support parametric cost estimates of unmanned, earth-orbiting space vehicles with a communications payload. For satellites with a mass less than 500 kilograms (kg), the SSCM is used. Because this thesis includes an analysis of various size satellites, the cost models were scaled to account for the variety of satellite sizes included in the scenario. Figure 3-4 illustrates the relationship of the objective functions to the MOGA. The cost model is highlighted in yellow to re-iterate that it is a historic model used as an estimate in this simulation.

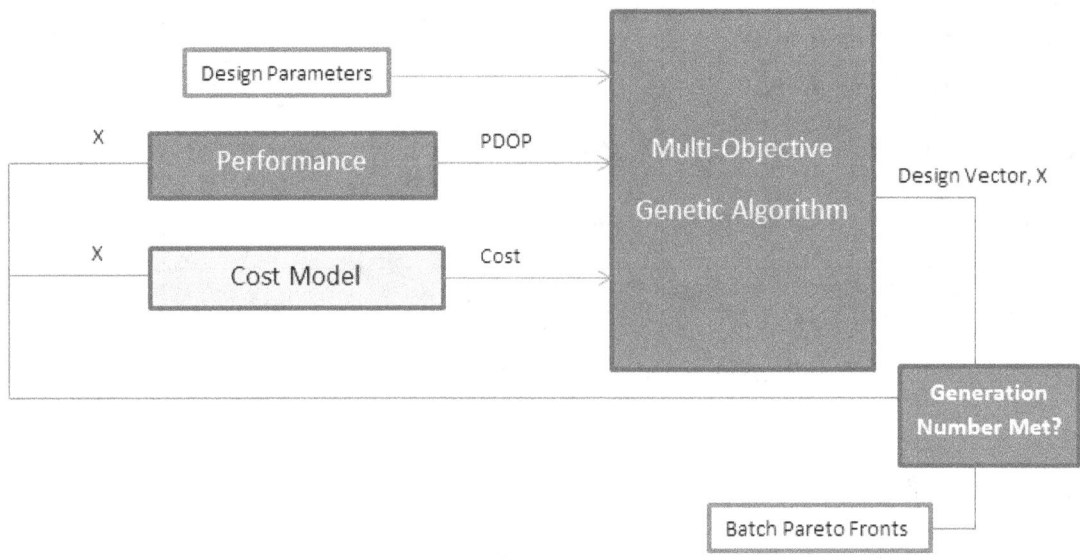

Figure 3-4: Simulation Structure

3.2 Constellation Analysis

This section describes the specific simulation parameters used to perform the analysis. When initializing the STK program, the scenario is set to 48 hours with a scenario step size of five seconds. The units for the scenario are set in kilometers. Once a satellite is created in STK, it is defined using certain properties. The classical elements are used to define the orbit, and the Two Body propagator is selected. The step size for this object is 60 seconds. For the coordinate type, Cartesian is chosen, and for the coordinate system, the International Celestial Reference Frame (ICRF) is used.

To analyze coverage, a coverage definition object is created in STK, and this object is set with a point granularity of six degrees. The coverage is measured using a figure of merit (FOM). The type of FOM is selected as dilution of precision, and it is computing the maximum. PDOP is used as the method, and the type is set as over-determined. The time step for this object is 300 seconds. For the MOGA, each test case may have a different number of generations and population size. The number of design variables for each satellite type is eleven including the fixed variables. Equation 12 represents the design vector, X. For the initial test cases, there is only one satellite type. For the hybrid constellation designs, there were two types of satellites.

$$X = \{[w_1, w_2, ..., w_{11}]_1, ..., [w_1, w_2, ..., w_{11}]_N\} \tag{12}$$

Where N is the number of satellite types.

3.3 Objective Function Calculations

Using the assumptions and parameters defined above, the objective functions are calculated. This section outlines the programming details for the objective functions. Each objective function is calculated in separate MATLAB files. Then in the wrapper function, called `navigation_gamultiobj.m`, the fitness function is defined as the concatenation of both objective functions. The function that calculates PDOP is called, `PDOP.m`, and the function that determines cost is `cost.m`.

3.3.1 PDOP.

The PDOP function accepts the design vector, X, and it outputs a global value of PDOP for that specific design. Within this function, STK is called using the Connect commands [35]. To improve computation time, the visibility for STK is set to zero, so the program does not display each time the PDOP function is called, but it runs in the background. A STK scenario is created using the function `navigation.m`. This function defines the scenario time and saves the scenario in a set location, which is determined by the user. The user must ensure that the path location in `PDOP.m` matches the path file specified in `navigation.m`. Within STK, a satellite constellation object is created and named NAVcon.

Since Walker constellations are used in this analysis, a seed satellite is generated and called NAVSat. Using the seed satellite, Walker constellations are generated using the specific Walker parameters in the X vector. To analyze the coverage provided by the constellation, a coverage definition object is created and called Performance. The coverage provided by the NAVcon constellation is measured with a FOM, which in this analysis is PDOP. STK stores the PDOP values by latitude, and a matrix is formed with the following columns: latitude, minimum, maximum, average, standard deviation, count, and sum. The maximum values are selected to analyze the worst case scenario for the specific design. Lastly, the median of the column of maximum values is used as the final value for PDOP.

3.3.2 Cost.

The cost function consists of the launch vehicle cost, USCM8 NRE, USCM8 recurring, and SSCM NRE. Equation 13 represents the total cost used in this thesis.

$$\text{Total Cost} = \text{Cost}_{LV/Plane} * N_{planes} + \text{Cost}_{NRE} * N_{dev} + \text{Cost}_{Recurring} * N_{prod} + \text{Cost}_{SSCM} * N_{prod} \quad (13)$$

Where

$$N_{Planes} = number\ of\ orbital\ planes$$

$$N_{prod} = total\ number\ of\ satellites$$

$$N_{dev} = 1\ when\ N_{prod} > 0$$

3.3.2.1 Launch Vehicle Cost.

This section details the calculations for launch vehicle cost. Before the launch cost can be calculated, the spacecraft mass must be determined. The spacecraft mass is calculated based on the mass of the payload. To develop a relationship between transmit power and payload mass, a second order polynomial trend is created using transmit power and payload mass of GPS[8], GLONASS[9], Galileo[10], and Beidou[11]. Table 3-3 illustrates the specific values used and Figure 3-5 shows the trend of the data. Equation 14 represents the payload mass equation generated from the trend-line in Figure **3-5**.

	Transmit Power, W	Payload Mass, kg
Beidou	52	400
GPS	50	347
Glonass	40	250
Galileo	25	112

Table 3-3: Mass and Power for Navigation Constellations

---

[8] GPS payload mass was estimated using the ratio of Glonass payload mass to total dry mass. The value of power is from [40].
[9] Glonass payload mass is from [39]. Its transmit power was estimated using the link budget equations.
[10] Galileo payload mass is from [38]. The transmit power is from [39].
[11] Beidou payload mass is from [35]. The transmit power was estimated using the link budget equations.

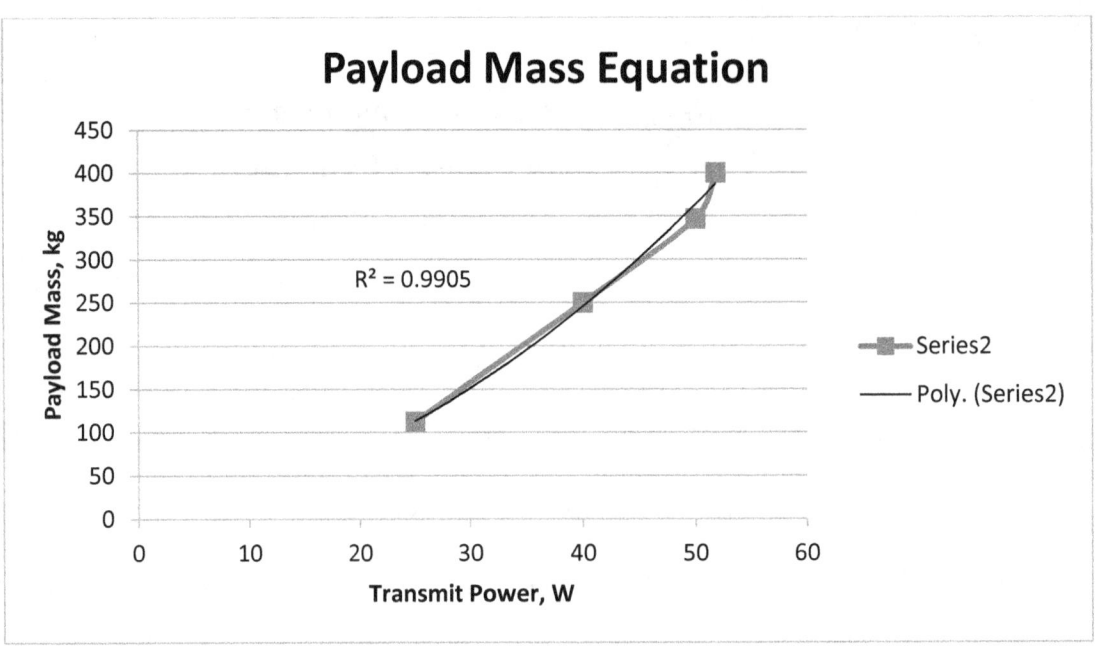

Figure 3-5: Payload Mass Equation

$$mass_{payload} = 0.116 * power_{tx}^2 + 1.279 * power_{tx} + 8.395 \qquad (14)$$

See Appendix B for detailed equations and calculations of estimated GPS payload mass

and transmit power values for GLONASS and Beidou. Equation 15 is from SME-SMAD

Table 14-18, and it represents the spacecraft dry mass.

$$Spacecraft_{dry_{mass}} = \frac{mass_{payload}}{0.32} \qquad (15)$$

The total spacecraft mass is then calculated using Equation 16.

$$Spacecraft_{mass} = mass_{payload} + Spacecraft_{drymass} \qquad (16)$$

The cost function accepts the design vector, X, along with a structure called lv.

The lv structure defines the different characteristics of the launch vehicles used in this

thesis. All launch vehicle values are from [7]. The structure consists of two fields: *mass*

and cost. The mass field possesses two fields: *bounds and alt.* The bounds field is a matrix of the launch capacities for each vehicle at LEO and GEO. The *alt* field is composed of 2000 km and infinity. This field is used to determine what altitude the scenario is using to provide the proper launch capacity values. The *cost* field is made up of *average* and *efficiency*. The average field is a matrix of the average launch cost for the different vehicles. The average launch cost includes the launch vehicle and related launch services [7]. The last field is a matrix of the cost efficiencies of the vehicles for LEO and GEO, which is measured as the cost per kilogram placed into orbit.

The launch vehicle analysis uses the spacecraft mass to determine which vehicle to use, how many vehicles to use, and launch cost. When the scenario is at LEO, the choices include Pegasus, Minotaur, Falcon 9, Atlas V, and Delta4H. Figure 3-6 illustrates the capacity and cost as the number of launch vehicles increases. The figure shows when certain vehicles may be more beneficial to use than others.

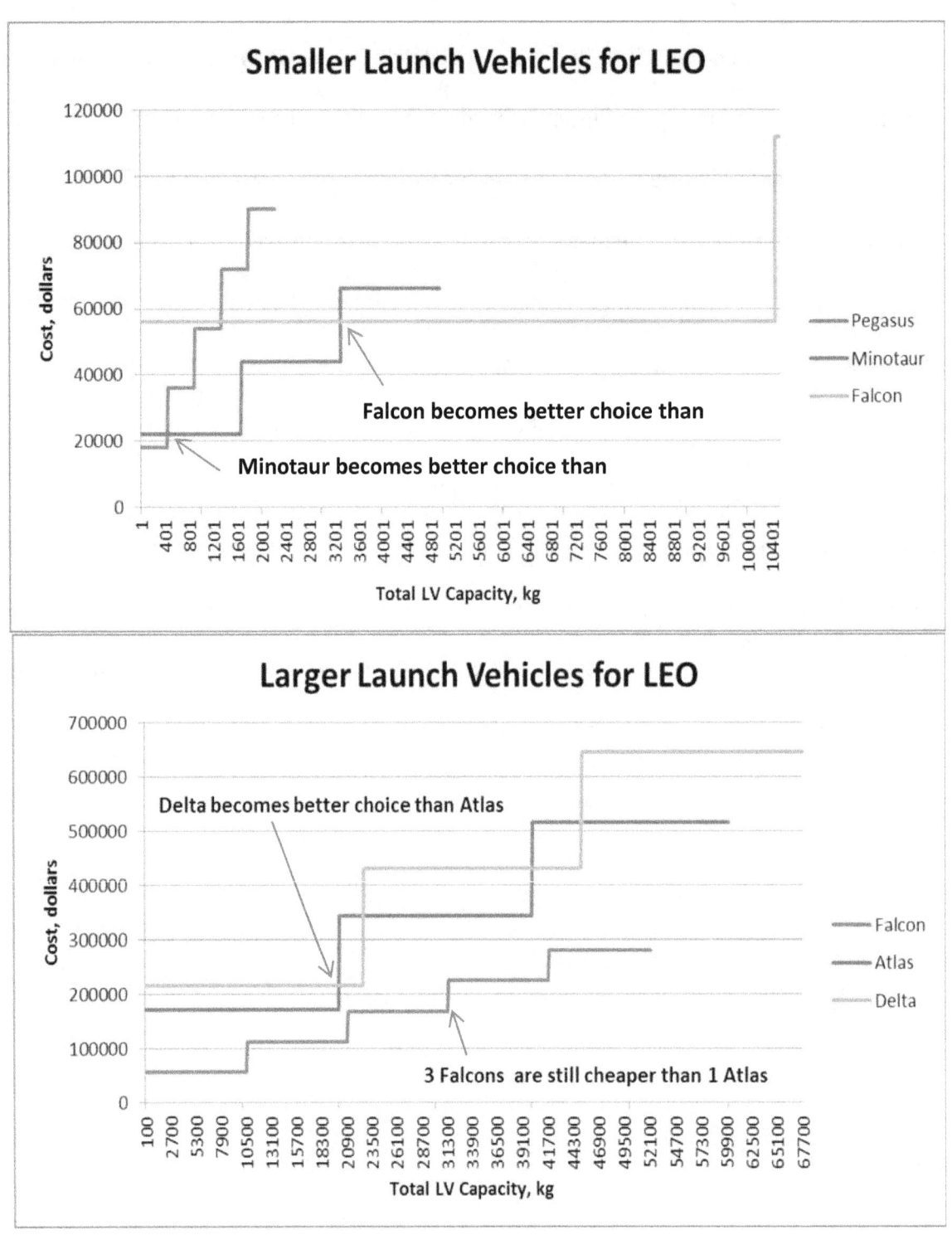

Figure 3-6: Launch Vehicles for LEO

When the scenario is at GEO, the available vehicles are Falcon 9, Atlas V, and Delta 4H.

Figure 3-7 demonstrates that three Falcons are cheaper than one Atlas or Delta.

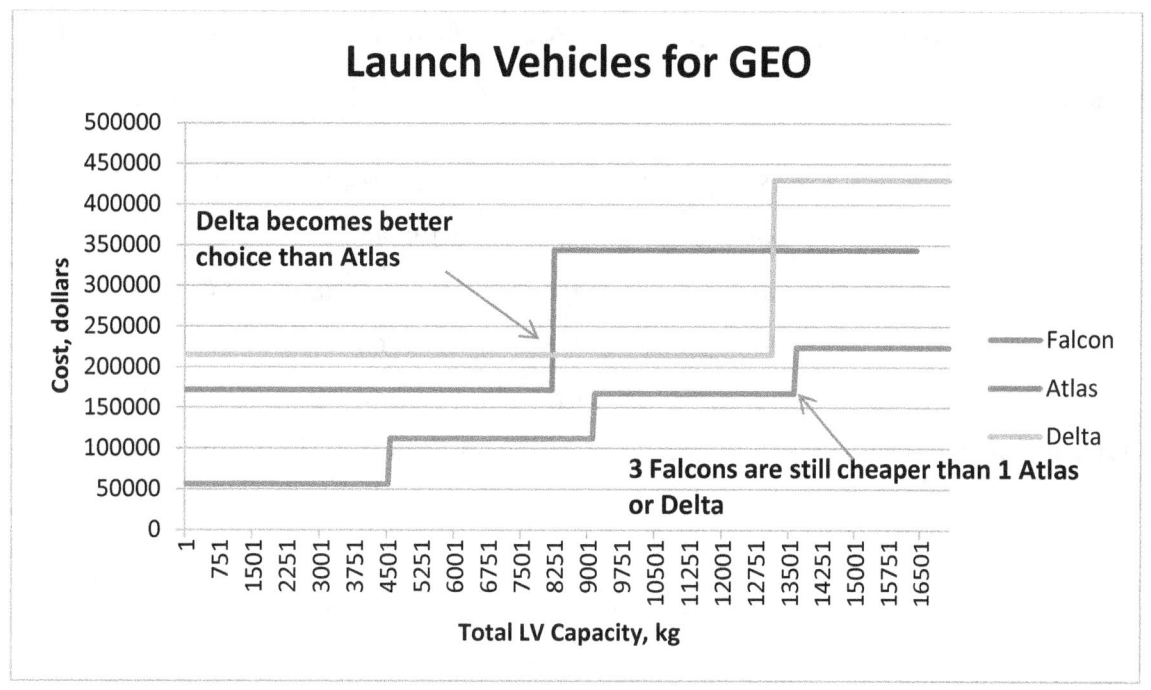

Figure 3-7: Launch Vehicles for GEO

After the spacecraft mass, number of launch vehicles, *n,* and the cost of those

vehicles are determined for one orbital plane, it is then multiplied by the number of

planes for total launch cost. Equation 17 represents the launch cost per plane, $Cost_{LV/Plane}$.

$$Cost_{LV/Plane} = \left[n_{Pegasus}, n_{Minotaur}, n_{Falcon}, n_{Atlas}, n_{delta} \right] * \begin{bmatrix} Cost_{Pegasus,} \\ Cost_{Minotaur} \\ Cost_{Falcon} \\ Cost_{Atlas} \\ Cost_{Delta} \end{bmatrix} \quad (17)$$

3.3.2.2 SME-SMAD Cost Models.

This section explains the specific equations used for the spacecraft cost. The cost

function, Equation 13, calculates cost in 2010 thousands of dollars for the entire system.

The three cost models included in the cost function can be seen in Table 3-4 [7]. Figure

3-8: Diagram of Cost Model is a diagram of the three cost models from SME-SMAD and how they are combined for this thesis. Equation 18 represents the USCM8 NRE model.

$$Cost_{NRE} = Cost_{s/c\_bus} + Cost_{Payload} + Cost_{IA\&T} + Cost_{program\_level} + Cost_{AGE} \qquad (18)$$

Where,

$$Cost_{s/c\_bus} = cost\ of\ the\ spacecraft\ bus$$
$$Cost_{Payload} = cost\ of\ the\ payload$$
$$Cost_{IA\&T} = cost\ of\ integration, assembly, and\ test$$
$$Cost_{program_{level}} = cost\ of\ the\ program\ level$$
$$Cost_{AGE} = cost\ of\ the\ aerospace\ ground\ equipment$$

Each component seen above is defined respectively in Table 3-4. Equation 19 illustrates the USCM8 recurring cost.

$$Cost_{recurring} = Cost_{s/c\_bus} + Cost_{Payload} + Cost_{IA\&T} + Cost_{program\_level} + Cost_{flight\_support} \quad (19)$$

Where, each component is defined respectively in Table 3-4, and the same notation is used from above. Equation 20 represents the SSCM NRE cost.

$$Cost_{SSCM} = Cost_{s/c\_bus} + Cost_{Payload} + Cost_{IA\&T} + Cost_{program\_level} + Cost_{flight\_support} + Cost_{AGE} \qquad (20)$$

Where, each component is defined respectively in Table 3-4.

SME-SMAD Cost Models, FY2010$			
USCM8 Non-recurring Subsystem CERs in FY2010 Thousands of Dollars			
Element	Equation	Variable	Reference
1 Spacecraft Bus	$110.2*X$	X=Spacecraft weight (kg)	Table 11-8
2 Payload	$618*X$	X= Communications subsystem weight (kg)	Table 11-8
3 Integration, Assembly, and Test	$0.195*X$	X=Spacecraft bus + Payload non-recurring cost ($K)	Table 11-8
4 Program Level	$0.414*X$	X=Space vehicle and IA&T non-recurring cost ($K)	Table 11-8
5 Aerospace Ground Equipment (AGE)	$0.421*X1^{0.907}*2.244^{X2}$	X1=Spacecraft bus non-recurring cost ($K) X2=0 for comm sats X2=1 for non-comm sats	Table 11-8
USCM8 Spacecraft Bus Recurring T1 CERs in FY2010 Thousands of Dollars			
Element	Equation	Variable	Reference
1 Spacecraft Bus	$289.5*X^{0.716}$	X=Spacecraft weight (kg)	Table 11-9
2 Payload	$189*X$	X= Communications payload weight (kg)	Table 11-9
3 Integration, Assembly, and Test	$0.124*X$	X=Spacecraft Bus + Payload Recurring Cost ($K)	Table 11-9
4 Program Level	$0.320*X$	X=Spacecraft Recurring Cost ($K)	Table 11-9
5 Flight Support	5850	-	Table 11-9
SSCM Total Non-recurring Cost (development plus one protoflight unit)			
Element	Equation	Variable	Reference
1 Spacecraft Bus	$1064+35.5*X^{1.261}$	X=Spacecraft weight (kg)	Table 11-11
2 Payload	$0.4*X$	X= Spacecraft Bus Total Cost ($K)	Table 11-11
3 Integration, Assembly, and Test	$0.139*X$	X=Spacecraft Bus Total Cost ($K)	Table 11-11
4 Program Level	$0.229*X$	X=Spacecraft Bus Total Cost ($K)	Table 11-11
5 Flight Support	$0.061*X$	X=Spacecraft Bus Total Cost ($K)	Table 11-11
6 Aerospace Ground Equipment (AGE)	$0.066*X$	X=Spacecraft Bus Total Cost ($K)	Table 11-11

Table 3-4: SME-SMAD Cost Models

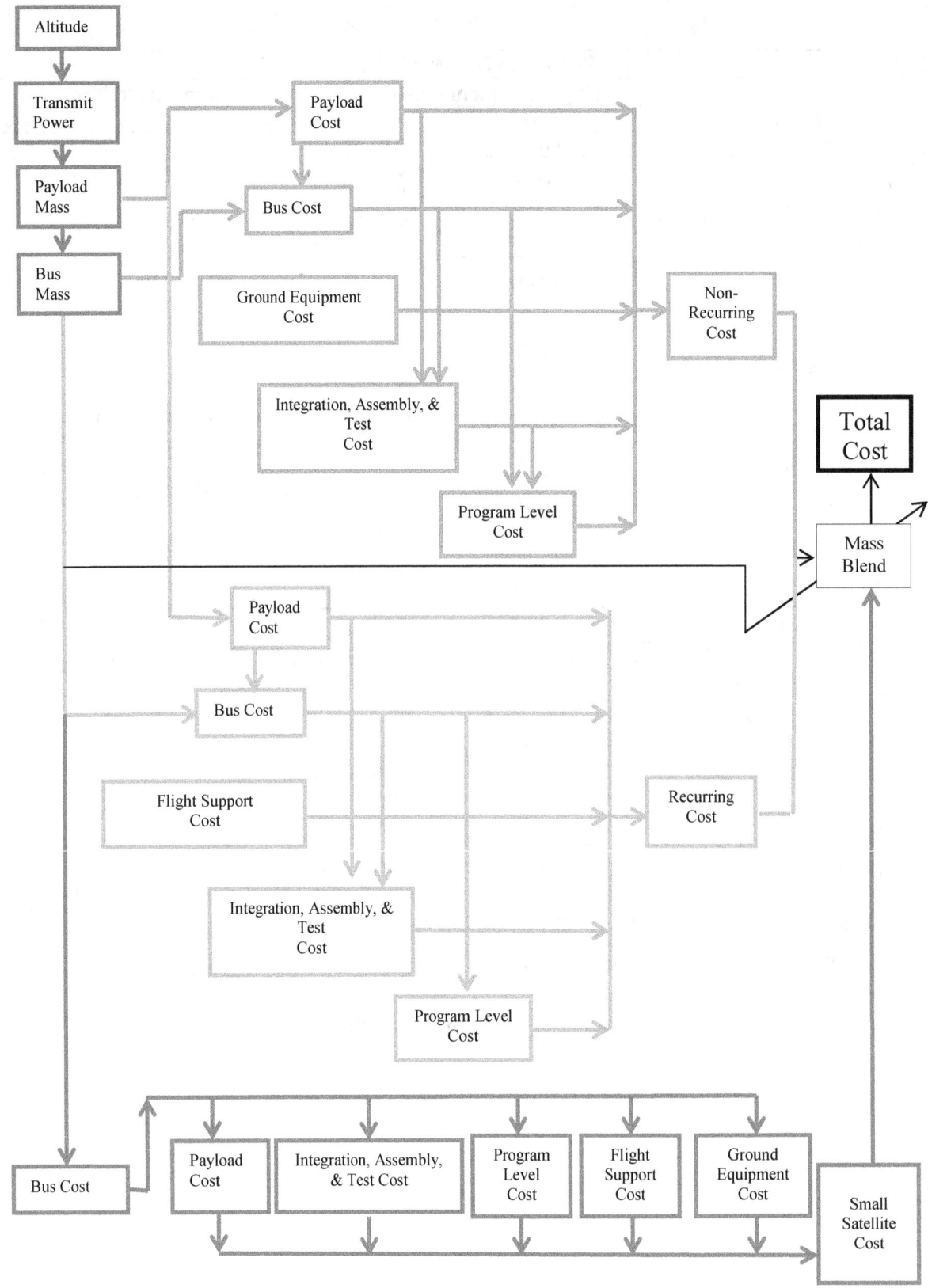

Figure 3-8: Diagram of Cost Model

3.4 Optimization Methodology

This section covers how the MOGA generates a population using the bounds on the design parameters and then eventually reaches batch Pareto fronts. To begin the optimization methodology, the proper bounds for the design parameters must be defined. An initial population is generated using the lower and upper bounds that are defined for each parameter. The algorithm creates a new generation using individuals in the current generation. The selection process is used to produce new populations. Once the stopping criterion is met, the algorithm stops. The stopping criterion used in this thesis is the total number of generations.

3.4.1 Population Size.

Population size and the number of generations are defined by the user. It is beneficial to create an initial population that provides a decent tradeoff between efficiency and effectiveness. The algorithm may not effectively search the space if the population size is too small, but if it is too large, the efficiency of the algorithm is significantly reduced to determine optimal solutions in a reasonable time of computation. The same logic applies for the generation size. For this thesis, the population size was 30 and the number of generations was 30.

3.4.2 Selection.

To create the new population, the algorithm scores each individual of the current population by computing its objective value. The raw fitness scores, (PDOP, cost) are scaled to convert them into a useful range of values. Based on their fitness, parents are selected from the population. The likelihood of selection is inversely proportional to how low their score is. The parents contribute their genes (entries of their vectors) to their children [17].

3.4.3 Mutation and Crossover.

Once the algorithm selects the parents from the population, the process to create children begins. Two operators were used: mutation and crossover. Mutation is one of the functions used to create children. Mutation children are created by randomly changing the genes of the individual parents. Mutation adds diversity to a population, and it increases the chances the algorithm will generate individuals with better fitness values. The crossover function is another function used to create children. This function randomly selects a vector entry from one of the two parents and assigns it to the corresponding entry for the child. Crossover allows the algorithm to take the best genes (parameters that result in low PDOP and cost) from different individuals and recombine them into superior children. In this thesis, the mutation and crossover functions specified that certain design parameters be integer values [17]. Equation 21 represents the variables that were integer values.

$$X = [w_1, w_2, w_5 \dots w_{1+(N-1)*11}, w_{2+(N-1)*11}, w_{5+(N-1)*11}] \qquad (21)$$

Where N is the number of satellite types.

3.4.5 Stopping Conditions.

The MOGA uses three different conditions to determine when to stop. Defining the total number of generations tells the algorithm to finish once the number of generations reaches the defined value. Another condition is when the average change in

3-70

the spread of the Pareto front is less than the user defined tolerance. The last condition is a time limit which, unless specified, is infinity. The reason for termination is output when the MOGA finishes [36] . For this thesis, the stopping criterion was the generation number.

3.5 Validation

To demonstrate that the design tool used in this thesis produces reasonable cost and performance values, the current GPS constellation is used to compare the values produced by the MOGA.

3.5.1 PDOP.

A separate MATLAB file, called `GPS_constellation.m`, was created to upload a GPS almanac to STK and generate the current GPS constellation. The same process from `PDOP.m` is used in this file to calculate a global value for PDOP. Because the median of the maximum values for PDOP is used as the objective function value for the MOGA, the same value was determined for the GPS constellation. STK calculated a PDOP value of 1.42 for the GPS constellation. This value is used to compare the PDOP values and design solutions generated from the MOGA.

To produce a design vector from the MOGA that matched the GPS, the design parameters were constrained to values close to those of the current GPS constellation. Table 3-1 illustrates the specific lower and upper bounds used in the validation case. The MOGA generated a Pareto front for these specific bounds. Figure 3-9 illustrates the Pareto front and shows the different number of planes and satellites per plane for the various solutions. The design solution that represents the current GPS is highlighted on the Pareto front. Table 3-5 shows the objective function values for the design solutions

generated for this case. Solution 3 is highlighted in green to show the relationship to the

GPS design solution on the Pareto front.

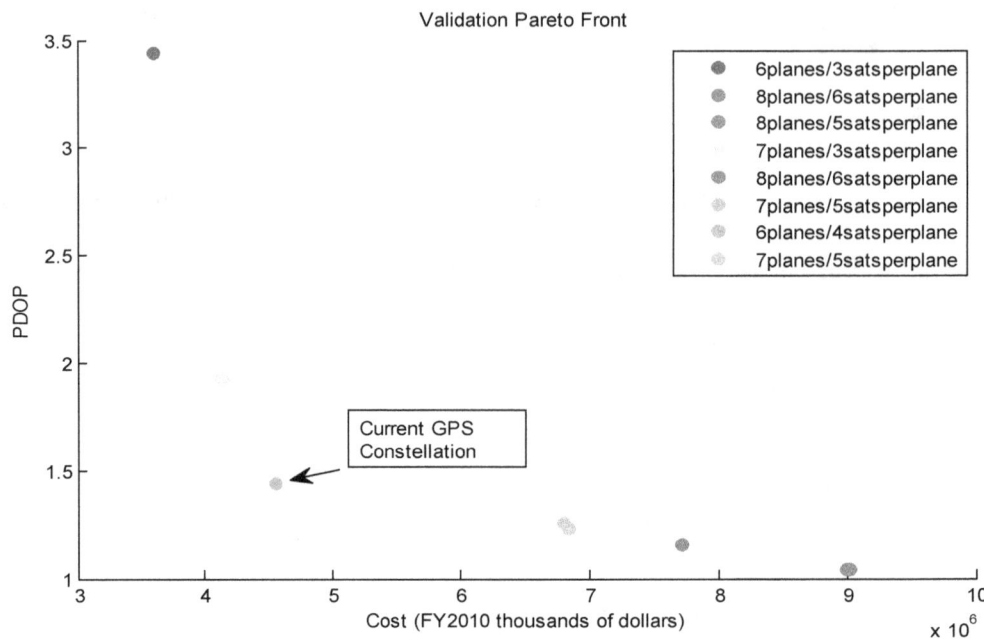

Figure 3-9: Validation Case

	Cost (FY2010 thousands of dollars)	PDOP
Solution 1	$ 3,587,470.35	3.44
Solution 2	$ 4,137,991.85	1.93
Solution 3	$ 4,555,058.80	1.44
Solution 4	$ 6,795,224.89	1.26
Solution 5	$ 6,837,865.44	1.23
Solution 6	$ 7,708,226.78	1.16
Solution 7	$ 8,987,404.19	1.05
Solution 8	$ 9,005,979.28	1.04

Table 3-5: Validation Results

Parameters	MOGA Results	GPS Values
#Planes	6	6
Sats/Plane	4	4
Truan, deg	5.4	-
RAAN Inc, deg	68.9	-
Alt, km	20,220	20200
Incl, deg	54.64	55.00
TX Power, W	48.90	50.00

Table 3-6: Validation Design Parameters

The MOGA successfully generated a design solution that roughly matched the current GPS constellation. Table 3-6 illustrates the design parameters generated by the MOGA for Solution 3. Section 2.2 discussed that the current GPS constellation has 6 orbital planes with four satellites per plane at an altitude of 20,200 km and is inclined to 55 degrees. The value for PDOP for Solution 3 is compared to the PDOP value generated for the current GPS constellation. Using the current GPS constellation PDOP value as the exact value, the percent error for the PDOP generated by the MOGA is 1.41%. These results validate that the design tool used in this thesis is capable of generating optimal navigation constellation design solutions.

3.5.2 Cost Function.

Since the current GPS constellation is used to compare the design tool results in this thesis to a real-world system, the cost value is compared with the cost of GPS Block III satellites. To compare the cost produced in the algorithm to the price per unit value for GPS Block III satellites, the launch cost was subtracted from the total cost generated by the MOGA. The total then is $4.21 billion. Using the unit cost of GPS Block III from [37], the cost generated by the MOGA is within 22% of that value.

3.6 Refining Results

Once the MOGA generates design solutions for the test cases used in this thesis, the results are refined. The lower and upper bounds for the design parameters need to be well defined for each different test case. For example, when analyzing results at 725 km, the amount of orbital planes and satellites per plane will be higher than a case at 20,000 km. If the design space defined for a certain test case is not well-suited, the Pareto front will show PDOP results much higher than the guideline set at six. There will also be a poor distribution in solutions.

When the results of a test case produce values for PDOP above 6, the results are refined by removing the invalid points and plotting the remaining points; however, the distribution in the points remains the same. For some cases, the distribution in solutions is poor. To improve this, the design space may need to be adjusted to adequately handle the specific test case. The bounds for certain design parameters may need to increase to allow for a larger design space. If a test case results in a design vector that possesses values for parameters that are equal to their corresponding lower bound value, it is concluded that decreasing the lower bound value to open the range for that parameter is beneficial to the MOGA. This method of refinement improves the design space for the MOGA and produces strong Pareto fronts for each test case.

3.7 Summary

This chapter outlined the design parameters, test cases, and equations used within the design tool. The objective functions and optimization process were discussed. The GPS constellation was used to compare the results of the design tool to a real-world system. The design solutions and analysis will be presented in Chapter 4.

Chapter 4
Results

The MOGA was used to analyze several different test cases based on altitude. Each test case required multiple runs to ensure valid results. LEO, MEO, and GEO altitudes were analyzed separately to determine possible design solutions (Section 4.1.1 thru Section 4.1.4). Hybrid constellations were analyzed as well. Trade-offs in each of the altitude ranges are discussed, and an analysis of the designs compared to the current GPS is given. Limitations of the work are also outlined in this chapter.

4.1 Results of the Simulation

The results of the simulation are separated based on altitude to demonstrate the initial design solutions. Each altitude range demonstrates different PDOP and cost values. The number of satellites required to obtain PDOP less than six decreases as the altitude increases. The cost values vary based on scenarios, but the general trend is that navigation constellations in LEO result in lower cost than MEO and GEO constellations.

4.1.1 LEO Results.

There were three test cases conducted within the LEO altitude range. Each test case resulted in Pareto fronts that demonstrated an improvement in performance for a higher cost. The LEO test cases had better results as altitude increased. The LEO results generated by the MOGA confirmed the expected trends between PDOP and cost, and they dominated the MEO results.

The results from the LEO altitudes were compared to determine the design solutions with the lowest PDOP and cost values. Figure 4-1 illustrates the three LEO

Pareto fronts together. The most desirable solution is the solution around the concave portion of the Pareto front. These solutions were chosen from the individual Pareto fronts rather than those seen in Figure 4-1 since the scaling in the comparison figure makes it difficult to see the concave portion. See Appendix C for the Pareto fronts of the individual test cases.

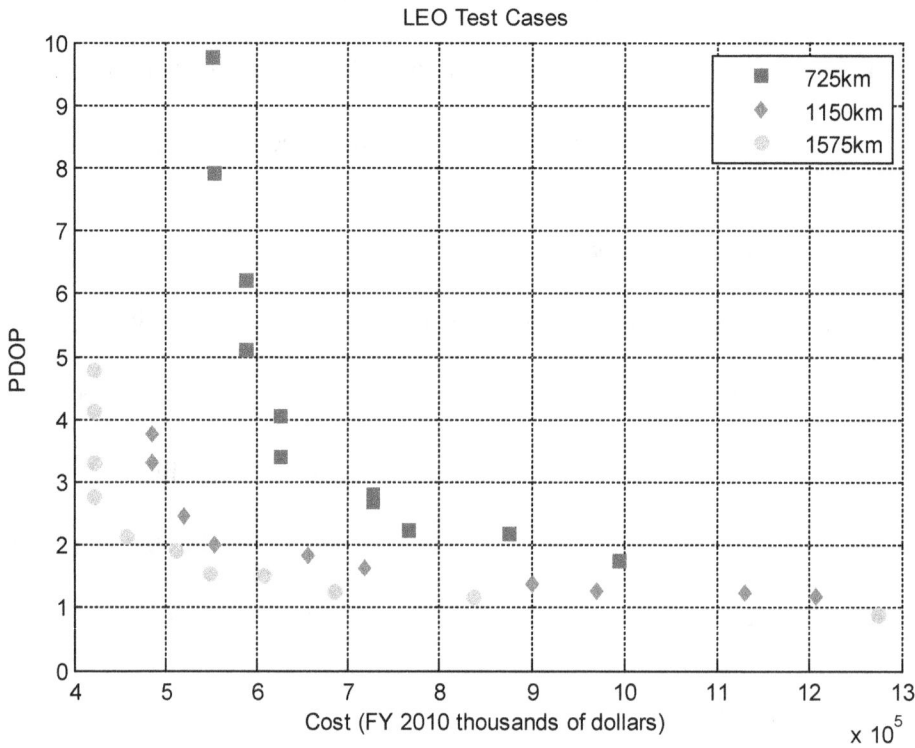

Figure 4-1: LEO Test Cases

The lowest altitude resulted in higher PDOP values and higher cost. This trend continued as the middle altitude had higher PDOP and cost values than the highest altitude. The points selected as the desirable solutions show that the highest altitude completely dominated the Pareto fronts at lower altitudes. Table 4-1 illustrates that when PDOP remains close to two, the cost decreases dramatically as altitude increases. Table 4-2 demonstrates that when cost remains relatively the same, PDOP decreases as altitude increases, but not as dramatically as the cost. Therefore, as altitude increases, there is more benefit in the cost than in the performance of the constellation designs produced within the LEO altitude range.

	PDOP	Cost ($B)
725 km	2	0.99
1150 km	2	0.55
1575 km	2	0.45

Table 4-1: Cost Tradeoff for LEO Test Cases

	PDOP	Cost ($B)
725 km	3.34	0.6
1150 km	1.82	0.6
1575 km	1.51	0.6

Table 4-2: PDOP Tradeoff for LEO Test Cases

4.1.2 MEO Results.

Within the MEO altitude range, there were four test cases that were analyzed. Each of the test cases illustrated an increase in cost for better PDOP values. The MEO results produced lower PDOP values than those in the LEO altitudes. The results verified the consistent trend between PDOP and cost.

To determine the difference in results for the MEO test cases, the four Pareto fronts were compared in Figure 4-2. The number of satellites required to produce the PDOP values decreased with altitude as seen in the LEO cases as well. Unlike the LEO test cases, there was a large difference in cost for the test cases in MEO when PDOP remained constant. The trend for the cost of the design solutions did not match the trend seen in LEO.

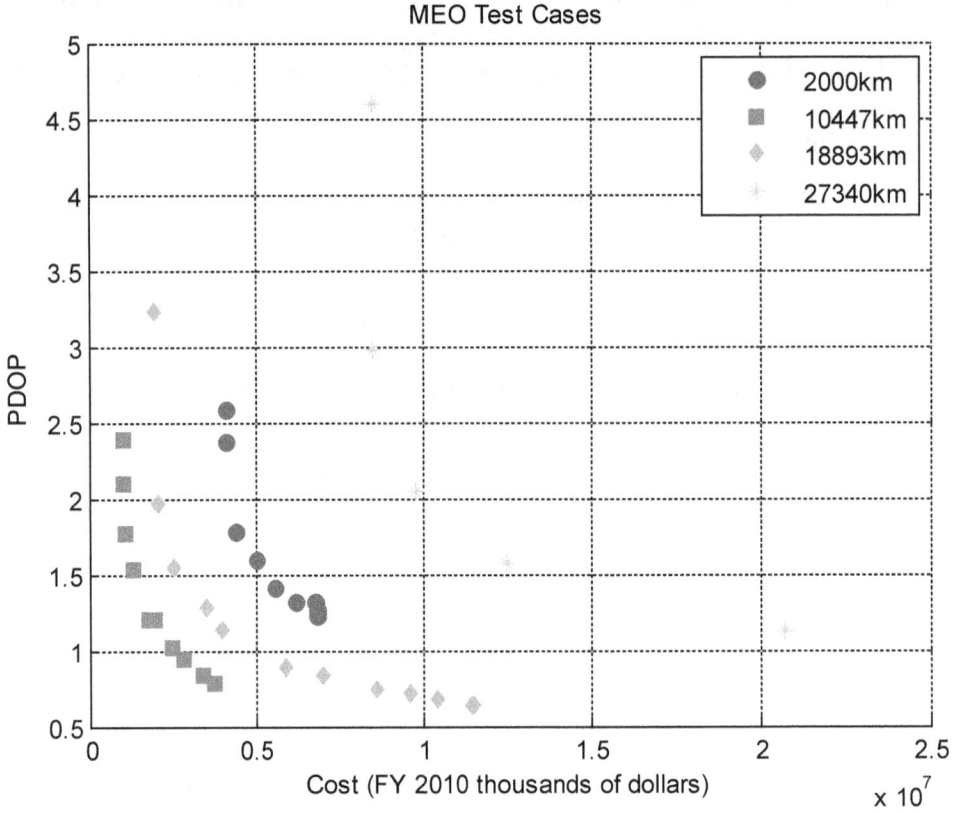

Figure 4-2: MEO Test Cases

When PDOP remains constant around 1.5, the last three altitudes show an increase in cost with an increase in altitude. The lowest altitude resulted in a higher cost than the middle two altitudes as seen in Table 4-3. The differences in cost as altitude increases are larger than in the LEO results. When the cost remains close to $3.9 billion dollars, Table 4-4 illustrates that the second MEO altitude possessed the lowest PDOP value. The highest altitude did not have a PDOP value related to that cost because the costs for this altitude were higher. The difference in PDOP as altitude increased remained similar to the differences in PDOP seen in the LEO results.

	PDOP	Cost ($B)
2000 km	1.5	4.9
10447 km	1.5	1.3
18893 km	1.5	2.5
27340 km	1.5	12.4

Table 4-3: Cost Tradeoff for MEO Test Cases

	PDOP	Cost ($B)
2000 km	2.37	3.9
10447 km	0.78	3.9
18893 km	1.14	3.9
27340 km	None	3.9

Table 4-4: PDOP Tradeoff for MEO Test Cases

To explain the unique trend in cost seen in the MEO range, the spacecraft and launch cost for one solution from each altitude test case was determined. The solution with a PDOP of around 1.5 was selected from each altitude for the comparison in cost. As a result of the larger step size in altitude used for the MEO range, there is a larger difference in the size of the satellites used in these designs. To determine whether the price is more quantity driven or size driven, Table 4-5 illustrates the difference in constellation cost values between each of the MEO cases.

The highest altitude, MEO case 4, resulted in the most expensive design as seen in Table 4-3. This is attributed to the large size in satellites required for that altitude. When compared to each of the other cases, the spacecraft cost for this altitude was the primary reason for the high cost values. MEO case 1, which is the lowest MEO altitude, possessed the next most expensive design. Table 4-5 shows that the price for this test case was driven by the large quantity of satellites required. This altitude required the most number of satellites to produce PDOP values less than six, and as a result, it required the most number of launch vehicles. MEO case 3 resulted in more expensive designs than MEO case 2, and this is driven by satellite size, as seen in Table 4-5. After

comparing the MEO results, MEO case 2 possessed the lowest PDOP value when cost

remained constant and the lowest cost when PDOP was set at 1.5.

MEO Test Case	Altitude (km)	PDOP	#Planes	# Satellites	# Launch Vehicles	Cost Per Satellite ($B)	Total Satellite Cost ($B)	Total Launch Cost ($B)
1	2000	1.5	8	88	8	0.42	36.80	0.45
2	10447	1.5	4	32	4	0.85	27.30	0.23
3	18893	1.5	5	25	5	1.92	48.00	0.28
4	27340	1.5	4	24	20	11.30	271.00	1.14

Table 4-5: MEO Cost Comparisons

4.1.3 GEO Results.

The last altitude range was at GEO. An altitude range of 35786-35796 km was

used for this test case. There were fewer designs solutions with PDOP values less than

six for this test case. Therefore, the Pareto front did not include as many solutions. The

solutions continued to follow the expected trend of an inverse relationship between

PDOP and cost.

4.1.4 Comparison.

Since there is such a large difference in cost between LEO, MEO, and GEO, only

the dominating Pareto front from LEO and MEO were used to compare with the GEO

designs. Figure 4-3 shows the comparison of the Pareto fronts for the different altitude

ranges.

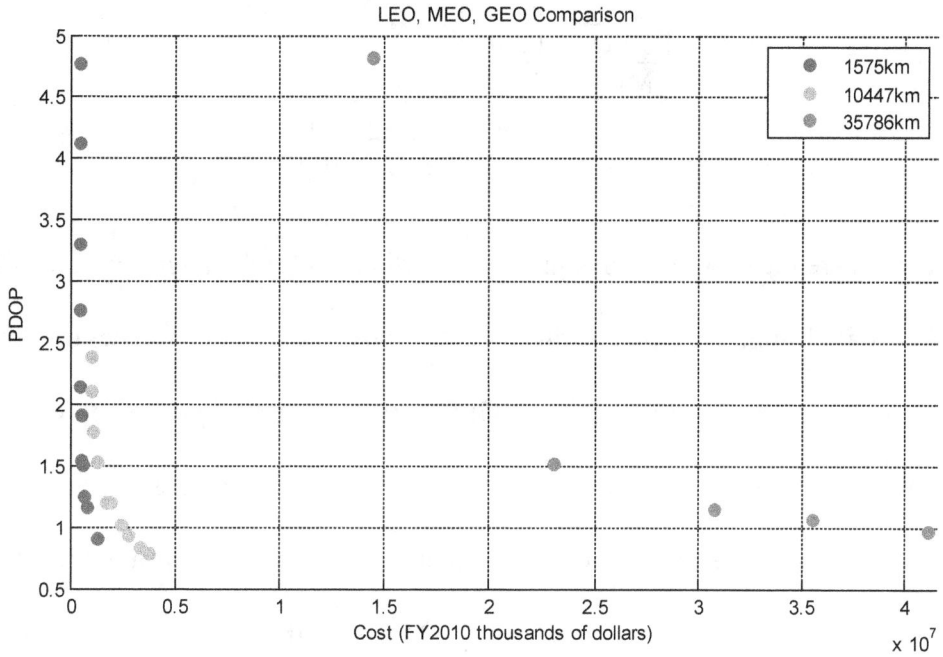

Figure 4-3: LEO, MEO, and GEO Comparison

The cost of the designs increases dramatically between each of the altitude ranges when PDOP is held constant around 1.5 (Table 4-6). The LEO design resulted in a lower cost than the MEO and GEO designs for a PDOP of 1.5. For a cost of about $1.3 billion, the LEO design resulted in a lower PDOP value than the MEO, and the GEO results did not produce a PDOP value for that price (Table 4-7). This price is the only price that resulted for LEO and MEO. There are no common prices between GEO and the lower altitudes.

	PDOP	Cost ($B)
LEO	1.5	0.61
MEO	1.5	1.31
GEO	1.5	23.10

Table 4-6: Cost Tradeoff for LEO, MEO, and GEO

	PDOP	Cost ($B)
LEO	0.90	1.3
MEO	1.53	1.3
GEO	None	1.3

Table 4-7: PDOP Tradeoff for LEO, MEO, and GEO

To compare the cost of the designs at the different altitude ranges, Figure 4-4 illustrates cost as altitude increases, and Figure 4-5 shows cost as the number of satellites increases. Figure 4-4 demonstrates that the GEO constellation designs are more expensive than the MEO and LEO designs. All but one of the MEO designs are more expensive than the LEO designs. There is one solution that overlaps with the LEO results. The number of satellites is greater in LEO than in GEO, but there are some MEO designs that possess the same number of satellites as LEO. For each altitude range, the cost increases as the number of satellites increases (Figure 4-5). The trend of altitude versus cost is affine for each of the altitudes.

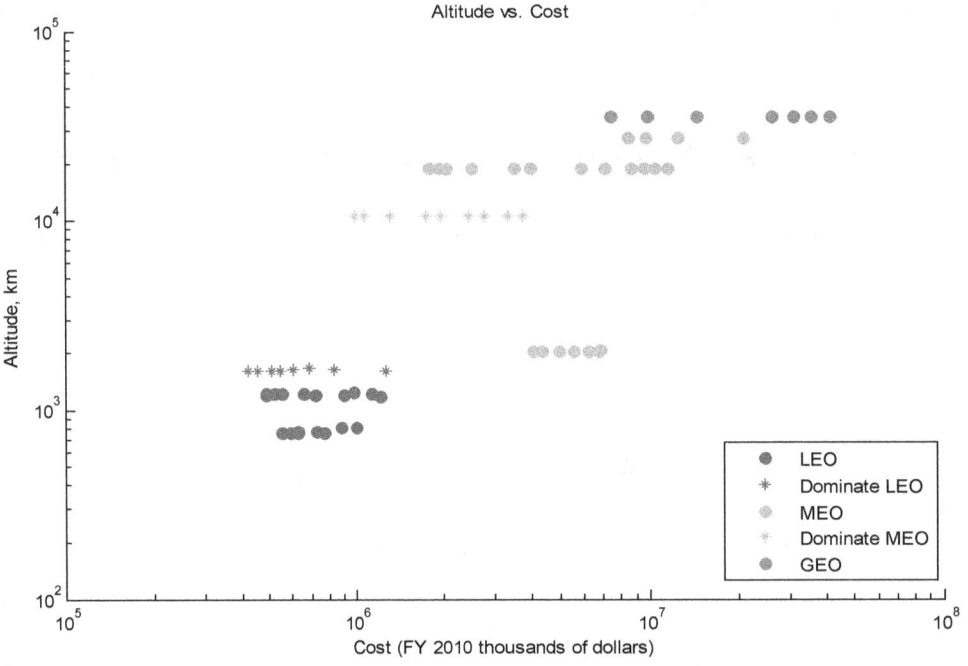

Figure 4-4: Altitude vs. Cost Comparison

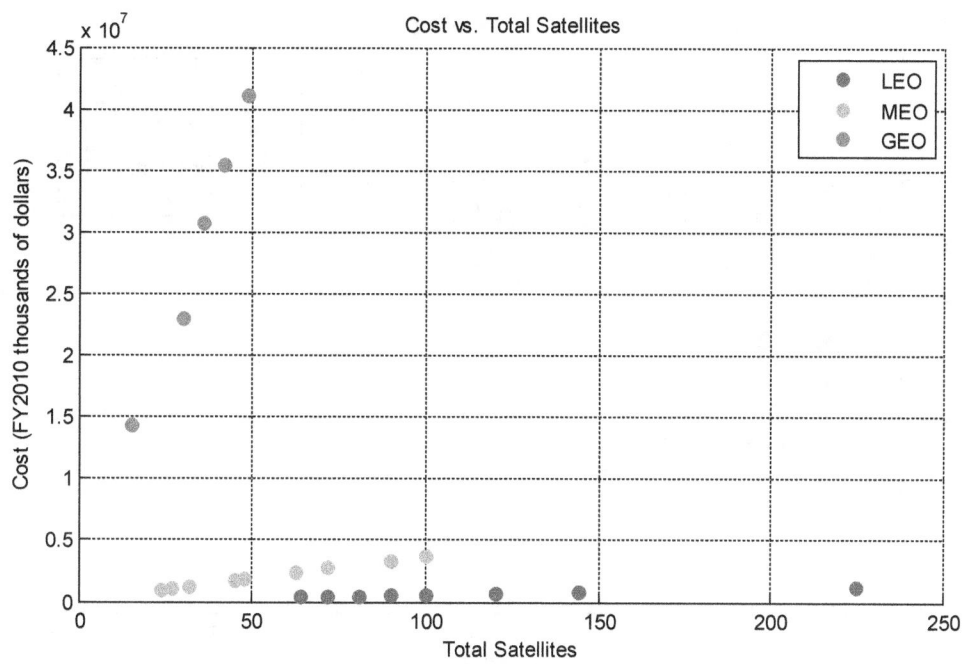

Figure 4-5: Cost vs. Total Satellites Comparison

PDOP was also analyzed for each altitude range as altitude and number of satellites increased. Figure 4-6 illustrates PDOP with respect to altitude, and it shows that the PDOP values improve as the altitude increases. All three altitude ranges produce similar PDOP values when PDOP is lower than two. Figure 4-7 shows the PDOP values as the number of satellites increases for each altitude range. The LEO designs had higher PDOP values than MEO and GEO. This figure also illustrates that the three altitude ranges begin to produce similar PDOP values, but LEO requires more satellites to reach those values.

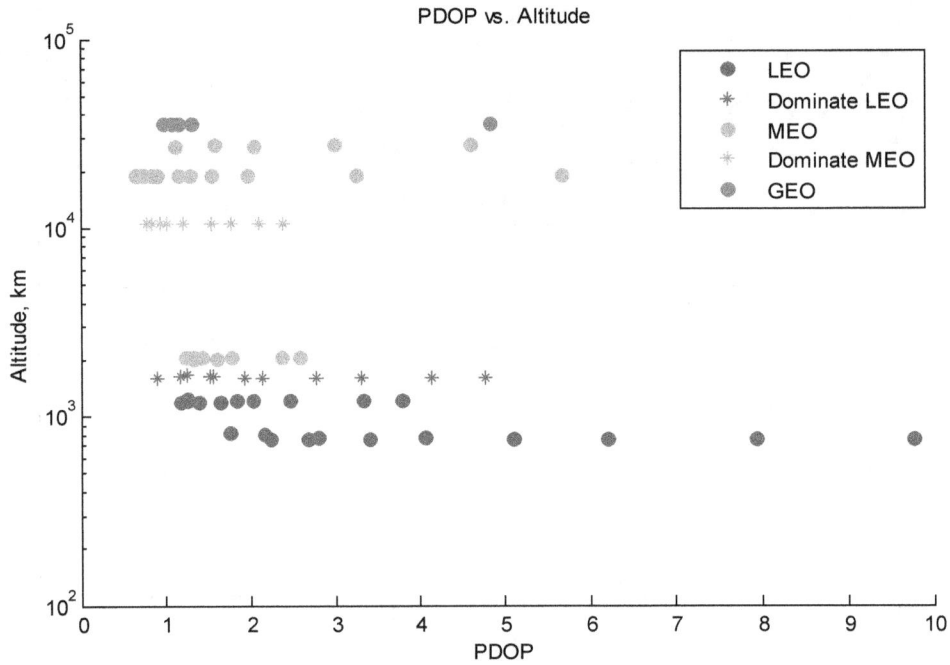

Figure 4-6: PDOP vs. Altitude

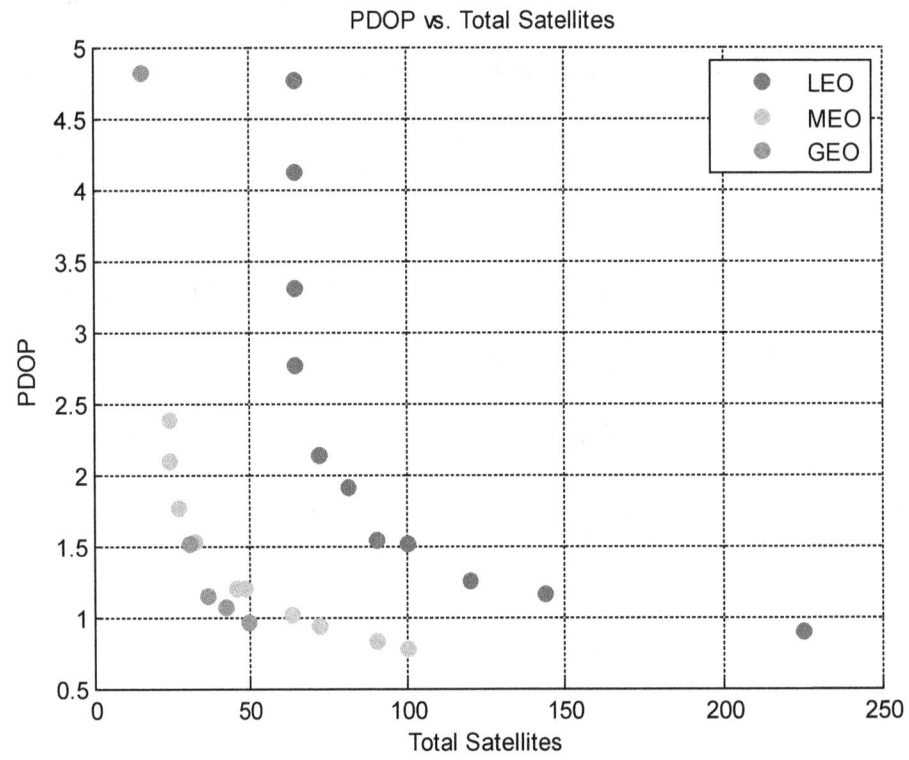

Figure 4-7: PDOP vs. Total Satellites Comparison

4.1.5 Hybrid Constellation Results.

The design tool used in this thesis has the ability to analyze multiple satellite

types. Therefore, three hybrid constellations were analyzed: a LEO-MEO hybrid, a LEO-

GEO hybrid, and a MEO-GEO hybrid. Figure 4-8 demonstrates the Pareto front from the

LEO-MEO hybrid designs with the individual altitude designs from LEO and MEO. The

hybrid constellation Pareto front fell very close to the MEO Pareto front.

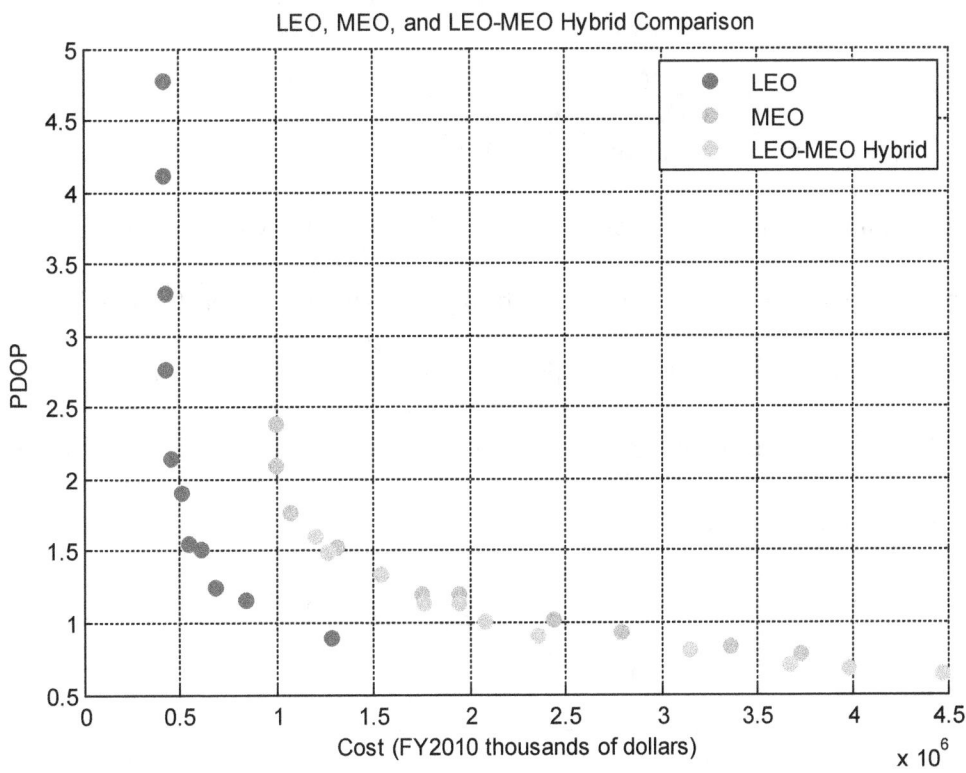

Figure 4-8: LEO, MEO, and LEO-MEO Hybrid Comparison

For a PDOP value of one, the LEO results still possessed a lower cost than the

MEO and the LEO-MEO hybrid. The hybrid cost value was much closer to the MEO

value than the LEO (Table 4-8). When setting the lower and upper bounds for the design

parameters for the hybrid test case, both altitudes possessed a lower bound of two for number of planes and number of satellites per plane. As a result, the MOGA did not have the option of selecting only MEO satellites or only LEO satellites for the constellations. The cost for the hybrid design is close to the individual MEO cost because the MOGA could select a LEO constellation design, but it was required to have at least four MEO satellites. This is a possible reason for the cost of the hybrid design being greater than the individual LEO designs.

	PDOP	Cost ($B)
LEO	1	1.27
MEO	1	2.08
LEO-MEO Hybrid	1	2.44

Table 4-8: Cost Tradeoff for LEO, MEO, and LEO-MEO Hybrid

At a constant price of $1.3 billion, the PDOP values showed an increase, with LEO possessing the lowest value and the hybrid possessing the highest value (Table 4-9). As demonstrated with the cost tradeoff, the PDOP value for the hybrid design was much closer to the MEO value. This demonstrates that there is no additional benefit in utilizing a LEO-MEO hybrid constellation.

	PDOP	Cost ($B)
LEO	0.90	1.3
MEO	1.49	1.3
LEO-MEO Hybrid	1.53	1.3

Table 4-9: PDOP Tradeoff for LEO, MEO, and LEO-MEO Hybrid

Figure 4-9 compares the trends of PDOP and total number of satellites for each altitude range along with the LEO-MEO hybrid designs. The hybrid design fell right between the LEO and MEO designs. The PDOP values were lower than LEO and MEO, but the number of satellites required was the same as MEO for PDOP greater than one and the same as LEO for PDOP less than one.

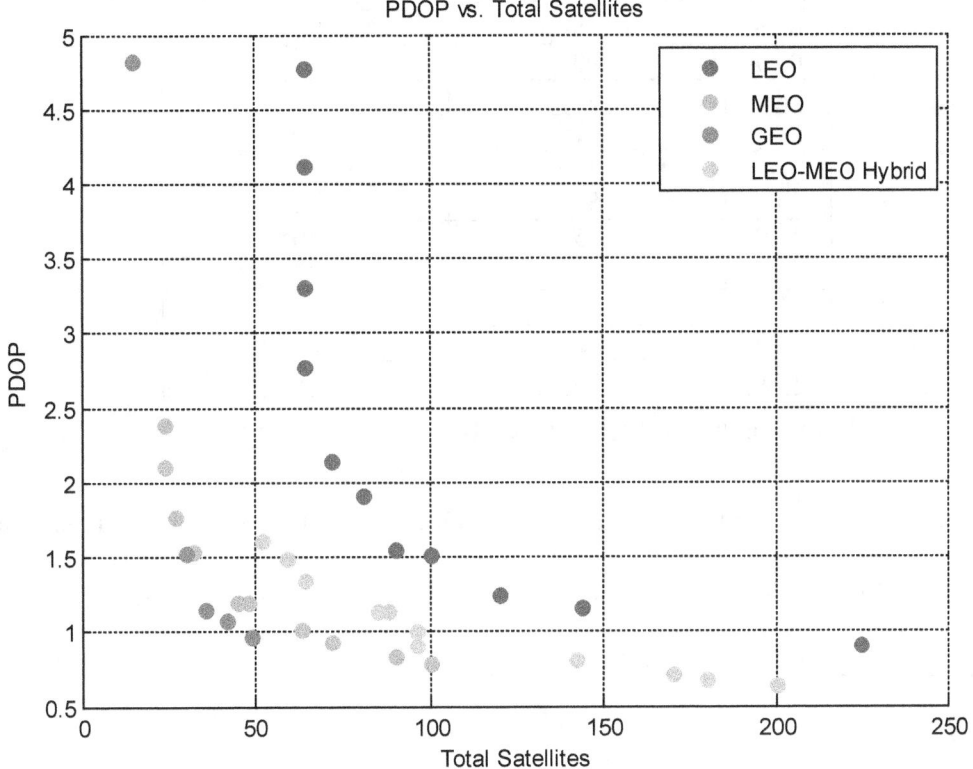

Figure 4-9: PDOP vs. Total Satellites for LEO-MEO Hybrid

Table 4-10 illustrates the separate number of LEO and MEO satellites within the hybrid designs. As PDOP increases, the number of satellites decreases, but for all but one design, there are the same number or fewer MEO than LEO satellites. This could be a result of the LEO satellites costing less than the MEO due to their smaller size.

4-87

PDOP	Total LEO Satellites	Total MEO Satellites
0.64	100	100
0.67	90	90
0.71	90	80
0.80	72	70
0.90	42	54
1.00	56	40
1.13	45	40
1.13	56	32
1.33	32	32
1.48	35	24
1.60	28	24

Table 4-10: Number of LEO and MEO Satellites in the Hybrid Designs

Figure 4-12 illustrates the individual LEO and GEO Pareto fronts compared to the LEO-GEO hybrid Pareto front. The hybrid front fell close to the GEO Pareto front, and showed some overlap in cost. Table 4-12 and Table 4-13 show the cost and PDOP tradeoffs.

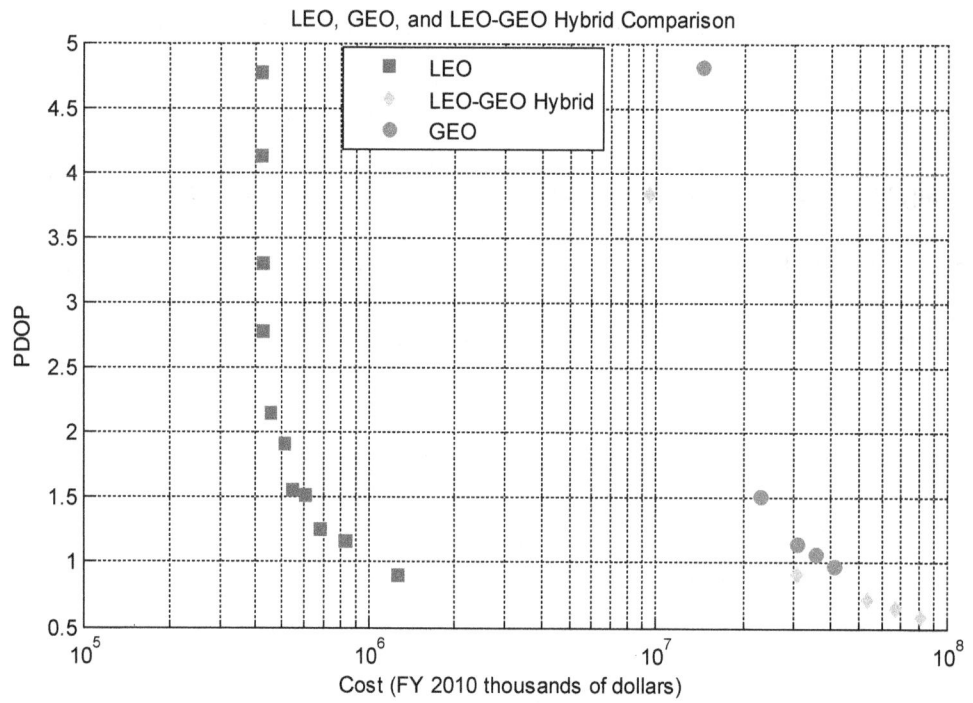

Table 4-11: LEO, GEO, and LEO-GEO Hybrid Comparison

When cost is analyzed for a PDOP of one, the individual LEO design has the lowest cost, and the individual GEO has the highest cost (Table 4-12). The LEO-GEO hybrid cost value was close to the cost value for GEO. This is attributed to the same issue mentioned for the LEO-MEO hybrid. The MOGA had to select satellites for both altitudes, so the cost would automatically be driven to higher costs than LEO.

	PDOP	Cost ($B)
LEO	1	1.27
GEO	1	41.13
LEO-GEO Hybrid	1	30.69

Table 4-12: Cost Tradeoff for LEO, GEO, and LEO-GEO Hybrid

When PDOP is analyzed for a constant cost of $30 billion, the LEO altitude does not have a design because the LEO designs have a much lower cost value. The LEO-MEO hybrid resulted in a slightly lower PDOP value than GEO. The cost and PDOP values for the hybrid designs remain closer to the higher individual altitude just as was shown for the LEO-MEO hybrid results.

	PDOP	Cost ($B)
LEO	None	30
GEO	1.16	30
LEO-GEO Hybrid	0.91	30

Table 4-13: PDOP Tradeoff for LEO, GEO, and LEO-GEO Hybrid

Figure 4-10 shows the change in the number of satellites for each altitude as the PDOP decreases. The hybrid design shows a similar number of satellites to LEO for a PDOP of close to one. Table 4-14 illustrates the number of satellites at the GEO altitude and LEO altitude within the hybrid designs. The number of GEO satellites is never greater than the number of LEO satellites.

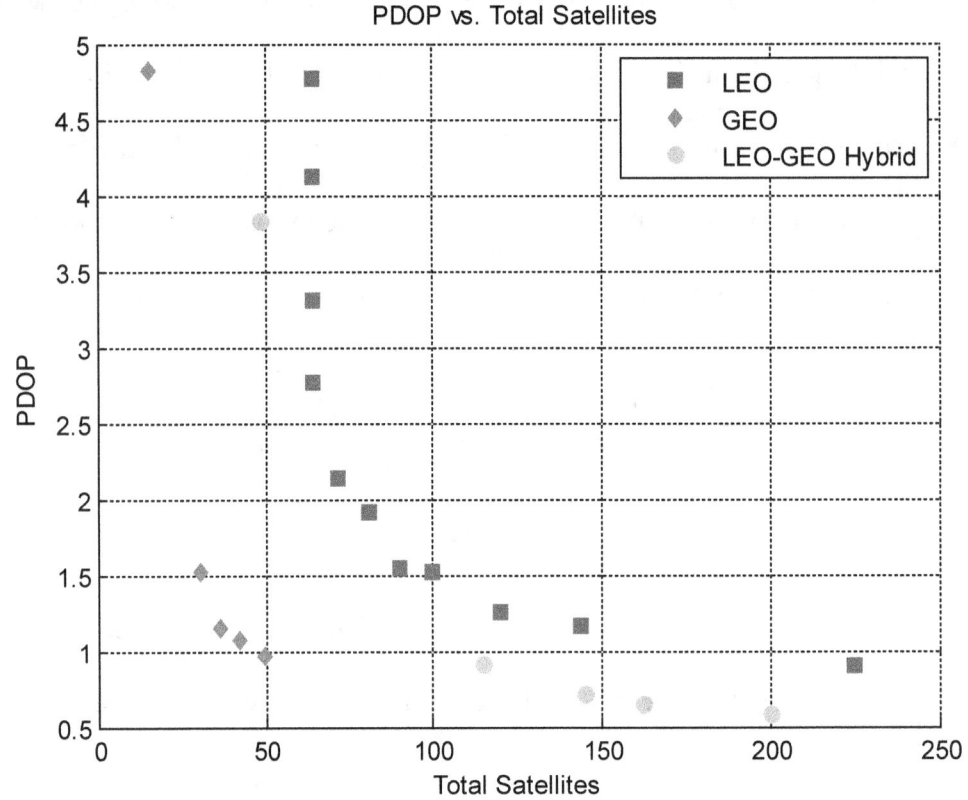

Figure 4-10: PDOP vs. Total Satellites for LEO, GEO, and LEO-GEO Hybrid

PDOP	Total LEO Satellites	Total GEO Satellites
0.58	100	100
0.65	81	81
0.71	81	64
0.90	80	35
3.83	40	8

Table 4-14: Number of LEO and GEO Satellites in the LEO-GEO Designs

The MEO-GEO hybrid constellation was compared to the individual MEO and GEO designs in Figure 4-11. The hybrid constellation's Pareto front fell between the MEO and GEO Pareto fronts. Table 4-15 and Table 4-16 demonstrate cost and PDOP tradeoffs.

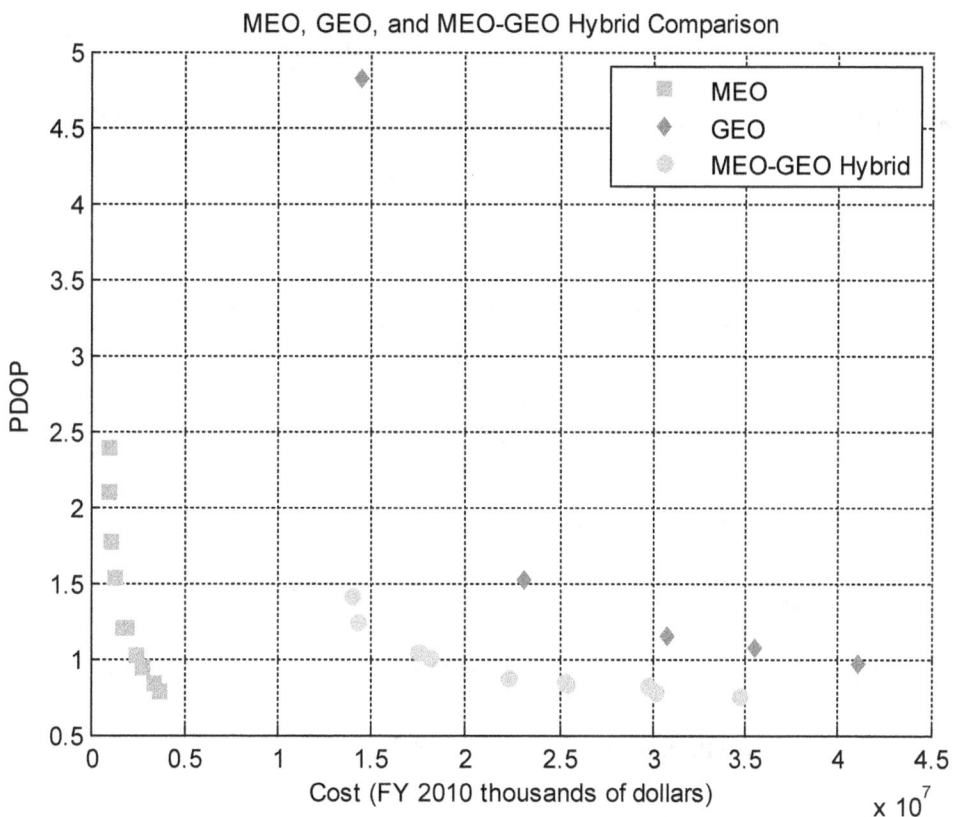

Figure 4-11: MEO, GEO, and MEO-GEO Hybrid Comparison

For a PDOP of one, the MEO results possessed the lowest cost, and the GEO results had the highest cost. There is a larger difference between the MEO-GEO cost and the GEO cost than was seen for the other hybrid results. The MEO-GEO cost is still much higher than the MEO results for the same reasoning used in the other hybrid cases.

	PDOP	Cost ($B)
MEO	1	2.44
GEO	1	41.13
MEO-GEO Hybrid	1	18.10

Table 4-15: Cost Tradeoff for MEO, GEO, and MEO-GEO Hybrid

For a cost of $30 billion, there is not a MEO design available because the MEO designs possess a much lower cost value. The MEO-GEO hybrid resulted in a lower PDOP value than the GEO results. They are still relatively close, as seen in the previous cases.

	PDOP	Cost ($B)
MEO	None	30
GEO	1.16	30
MEO-GEO Hybrid	0.79	30

Table 4-16: PDOP Tradeoff for MEO, GEO, and MEO-GEO Hybrid

The PDOP values as the number of satellites increased was analyzed for this hybrid constellation in Figure 4-12. For a PDOP of one, the MEO results show a larger number of satellites than the GEO and MEO-GEO results. The GEO and MEO-GEO results require the same number of satellites. Table 4-17 shows the number of satellites for MEO and GEO within the hybrid designs. The same trend continues, where there are fewer satellites in the higher altitude range for the hybrid designs.

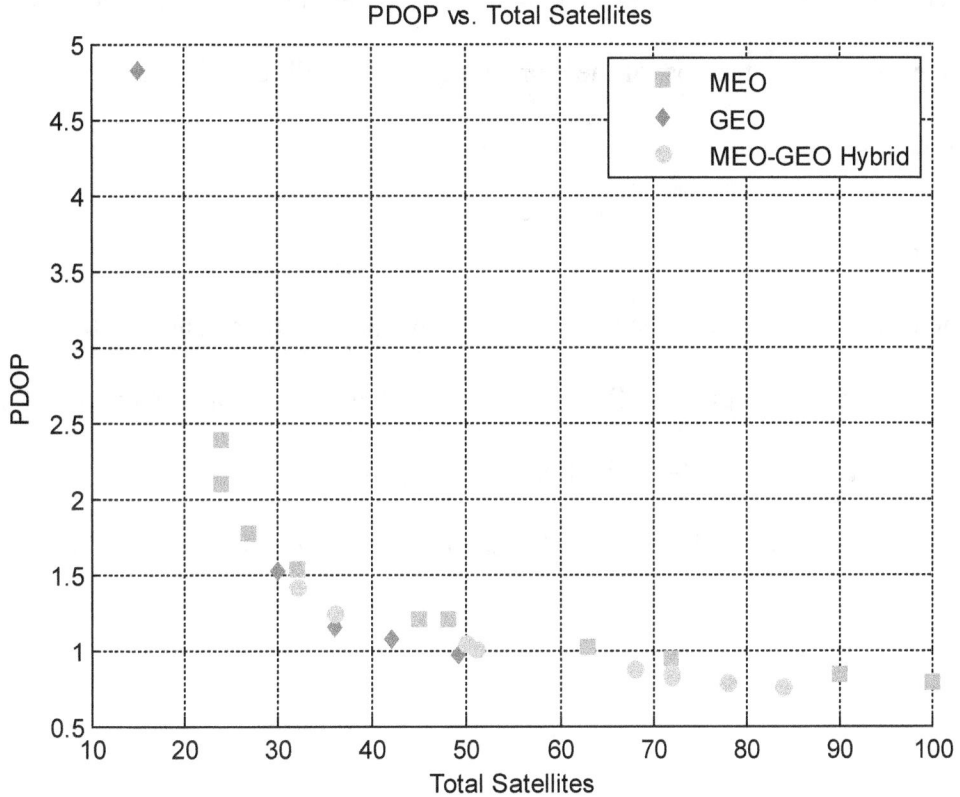

Figure 4-12: PDOP vs. Total Satellites for MEO-GEO Hybrid

PDOP	Total MEO Satellites	Total GEO Satellites
0.75	48	30
0.75	24	12
0.78	35	16
0.82	20	12
0.83	48	24
0.84	35	15
0.87	42	30
1.00	48	24
1.05	48	36
1.24	48	36
1.41	48	20

Table 4-17: Number of MEO and GEO Satellites in the MEO-GEO Designs

For each of the hybrid constellations that were analyzed, they produced similar

PDOP values to the lower altitude designs, but required a much higher cost. Therefore,

the hybrid constellations offered no benefit when compared to the individual altitude designs. This could also be a result of the hybrid constellations requiring satellites at each altitude. Therefore, the cost of the more expensive (satellites at higher altitude) satellites could not be removed.

4.2 Trade-offs

The results from each of the test cases demonstrated similar trends for the number of planes and satellites per plane as PDOP decreased. The inclination remained within a set range for most of the altitudes. The following graphs are separated into LEO trends and MEO and GEO trends. The MEO and GEO results were combined since they were relatively similar.

Figure 4-13 illustrates PDOP as the number of planes increases for the LEO altitudes. As expected, the PDOP values decrease as the number of planes increases. For the same number of planes, there are multiple points that show a decrease in PDOP. This is a result of some of the solutions possessing more satellites per plane. For a PDOP of about two, the number of planes decreases as the altitude increases.

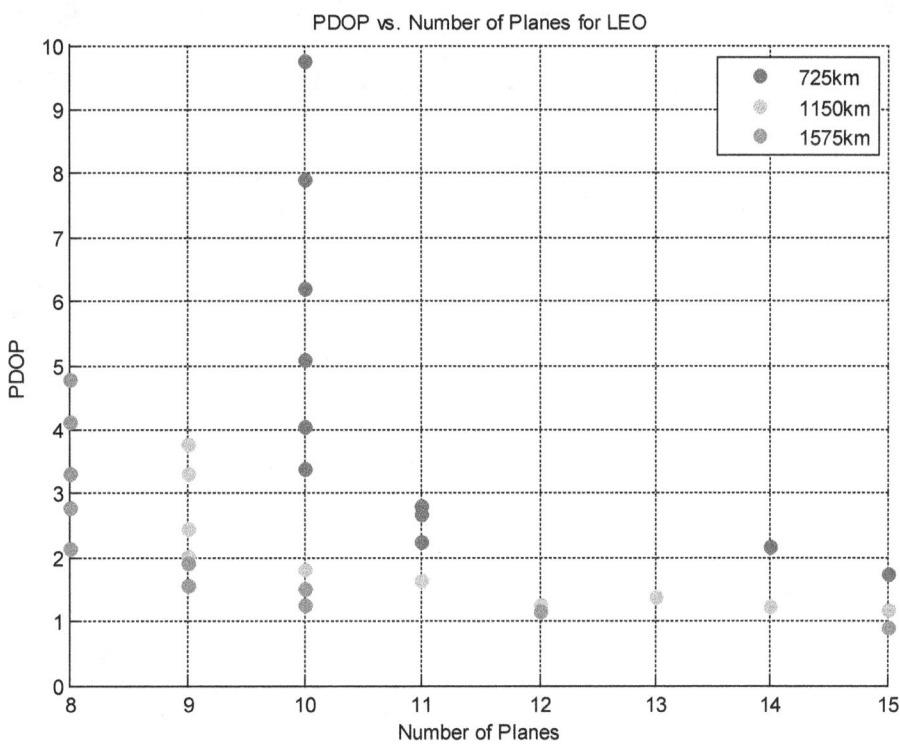

Figure 4-13: PDOP vs. Number of Planes for LEO

In Figure 4-14 the PDOP decreases as the number of satellites per plane increases for the LEO alitudes. This is the same trend seen with the number of planes. This makes sense due to the effect the total number of satellites had on PDOP (Section 4.1.4). For the same number of satellites per plane, there are points that show a decrease in PDOP. This is attributed to the points possessing more orbital planes. Some of the solutions possess the same number of planes and satellites per plane, but still demonstrate a change in PDOP. This is analyzed later with the effects of inclination.

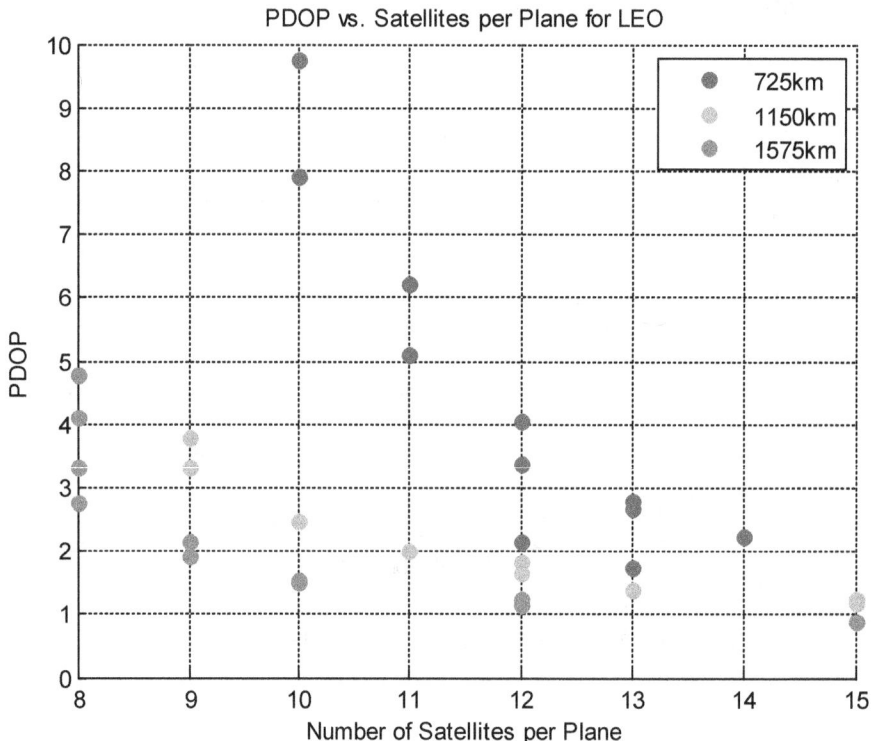

Figure 4-14: PDOP vs. Satellites per Plane for LEO

Figure 4-15 demonstrates the relationship between the number of satellites per plane and the number of planes. The difference between the number of planes and the number of satellites per plane is never greater than three. For the upper LEO altitude, the trend between satellites per plane and planes begins with a staircase effect. As the number of planes increases for each altitude, the number of satellites per plane either remains the same or increases as well.

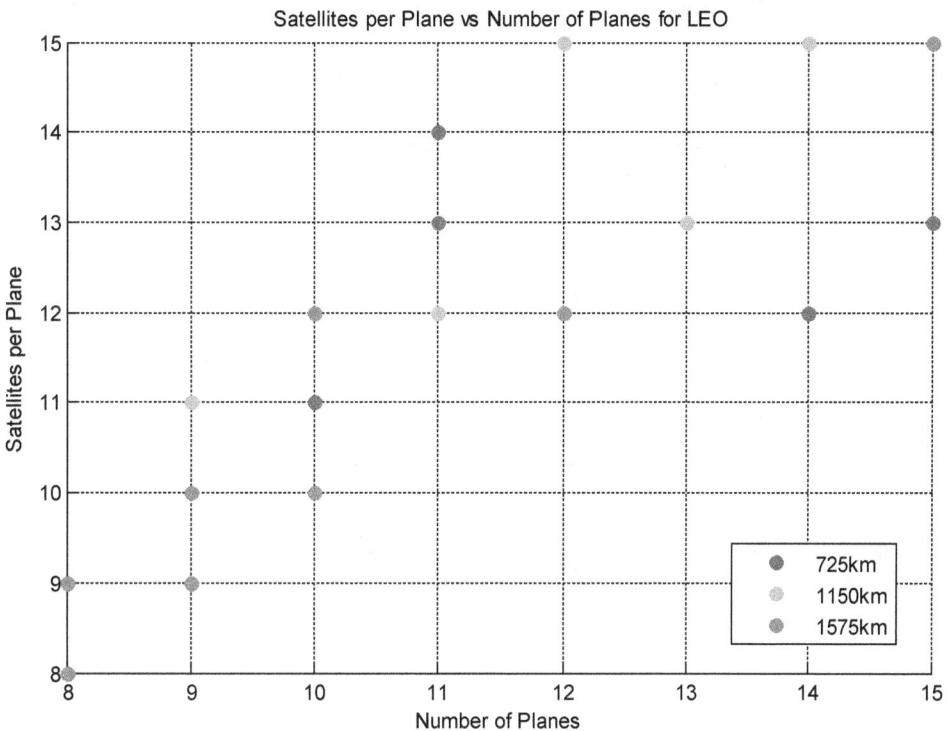

Figure 4-15: Satellites per Plane vs. Number of Planes for LEO

The third LEO test case, the test case that resulted in the dominating Pareto front in LEO, illustrates a roughly linear trend between the number of satellite per plane and the number of planes. Equation 22 represents the trend.

$$spp = 0.9757 * planes + 0.5951 \qquad (22)$$

Where $R^2 = 0.91$.

The inclination results did not demonstrate the same trend as seen with the previous two parameters. The inclination did not decrease or increase with PDOP for the lower LEO altitudes, but for the upper LEO altitude, the inclination showed an increase with an increase in PDOP. This is the only case where inclination showed this specific trend. For all the LEO altitudes, the inclination remained within a 10 degree range. For 725 km, the inclination was within 54 and 58 degrees. The results at 1150 km were within 48 and 58 degrees, and lastly, for 1575 km, the inclination remained between 36 and 46 degrees. For a PDOP of about two, the inclination is roughly equal for the lower LEO altitudes, but the upper LEO altitude had the lowest inclination.

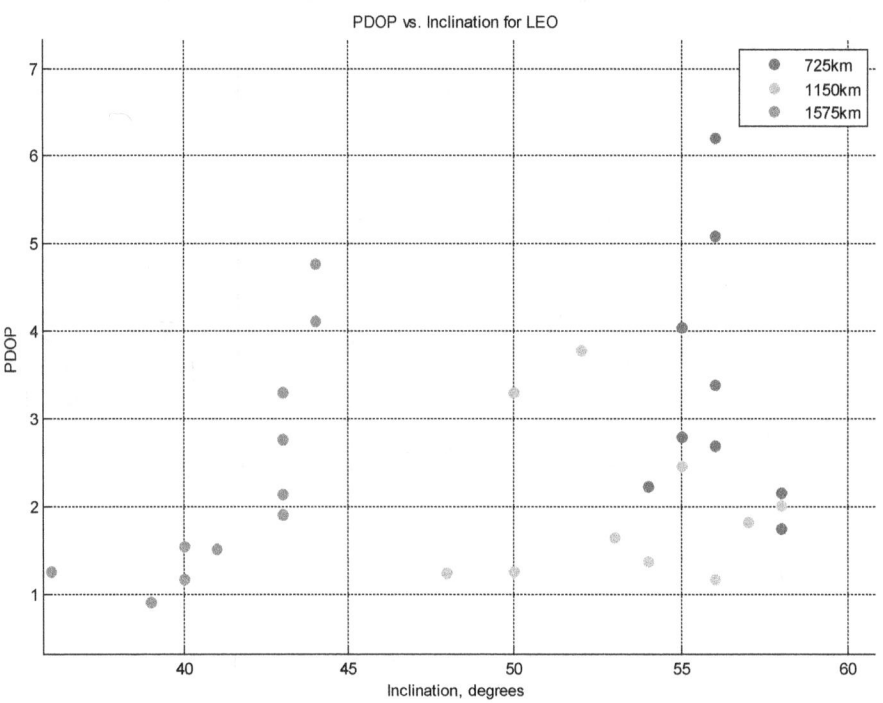

Figure 4-16: PDOP vs. Inclination for LEO

Since each LEO altitude possessed some solutions with the same number of planes and satellites per plane with a change in PDOP, the inclination was analyzed for these cases. Inclination is the only other design parameter than could affect PDOP if the number of planes and satellites per plane were constant. Figure 4-17 illustrates the solutions from the LEO altitudes with equal number of planes and satellites per plane, but resulted in different PDOP values. For each of the solutions, the inclination is slightly greater for the solution with a lower PDOP value.

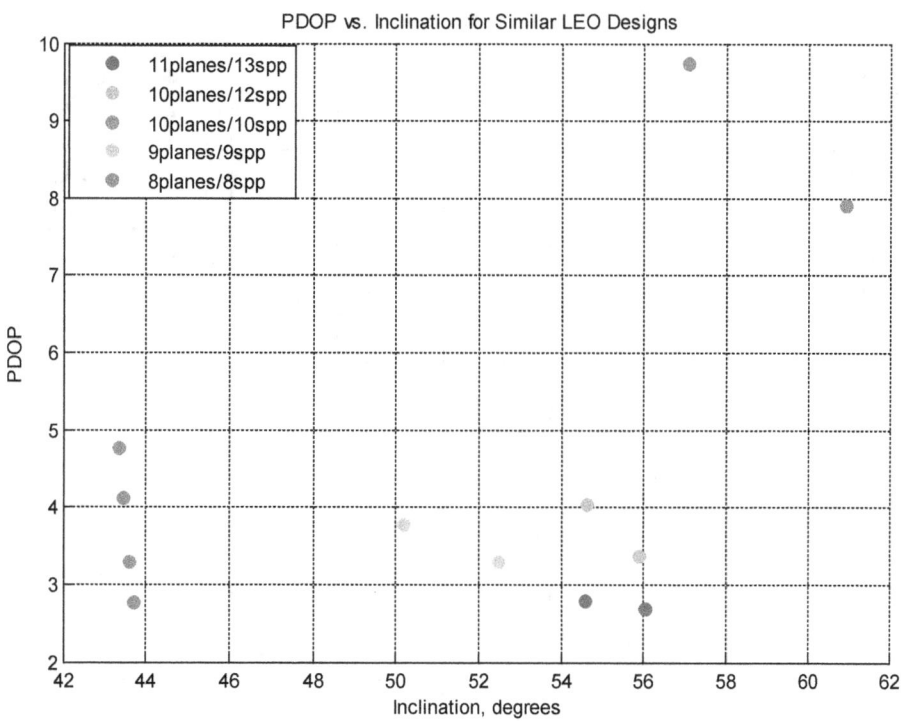

Figure 4-17: PDOP vs. Inclination for Similar LEO Designs

The same design parameters were analyzed for MEO and GEO. The number of planes demonstrated a consistent trend when compared to PDOP. This is shown in Figure 4-18. There was more of a variety in the number of planes for the MEO and GEO results, which illustrated a smoother curve as PDOP decreased. These results also demonstrate points with the same number of planes and a change in PDOP. This is consistent with the LEO results, and it is a result of a different number of satellites per plane for those solutions.

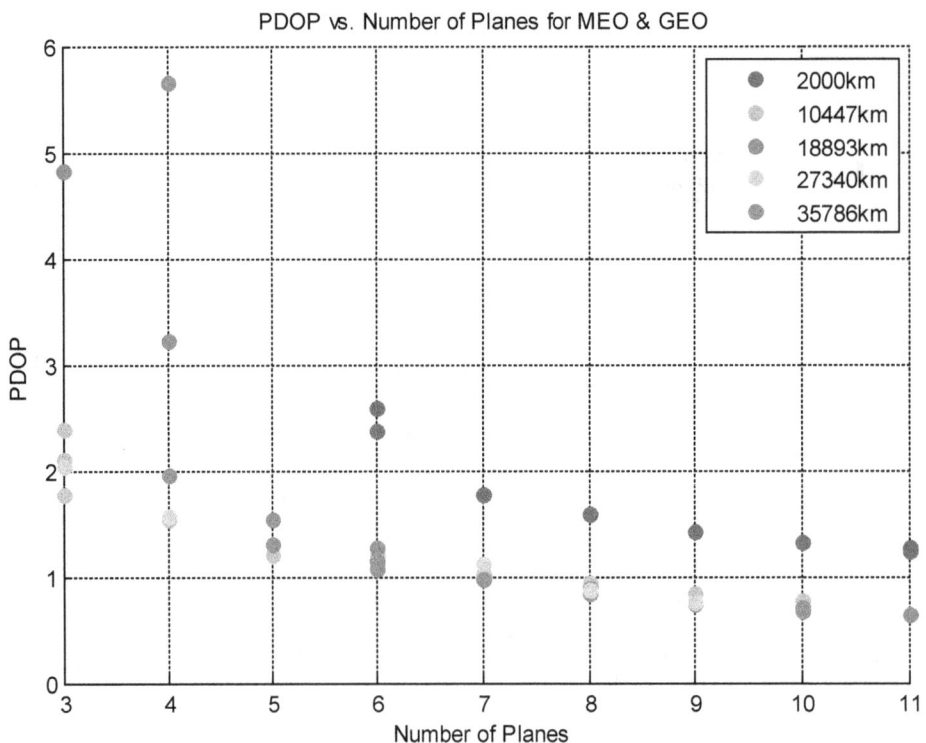

Figure 4-18: PDOP vs. Number of Planes for MEO & GEO

Figure 4-19 illustrates the number of satellites per plane as PDOP decreases for MEO and GEO. There is less of a variety in the number satellites per plane for these altitudes than were seen for the LEO results. Therefore, the trend is not as smooth. The solutions with the same number satellites but a change in PDOP could have a different number of planes. For a PDOP of about one, the number of satellites decreases as the altitude increases, and for 10447 km and 18893 km, the number is equal.

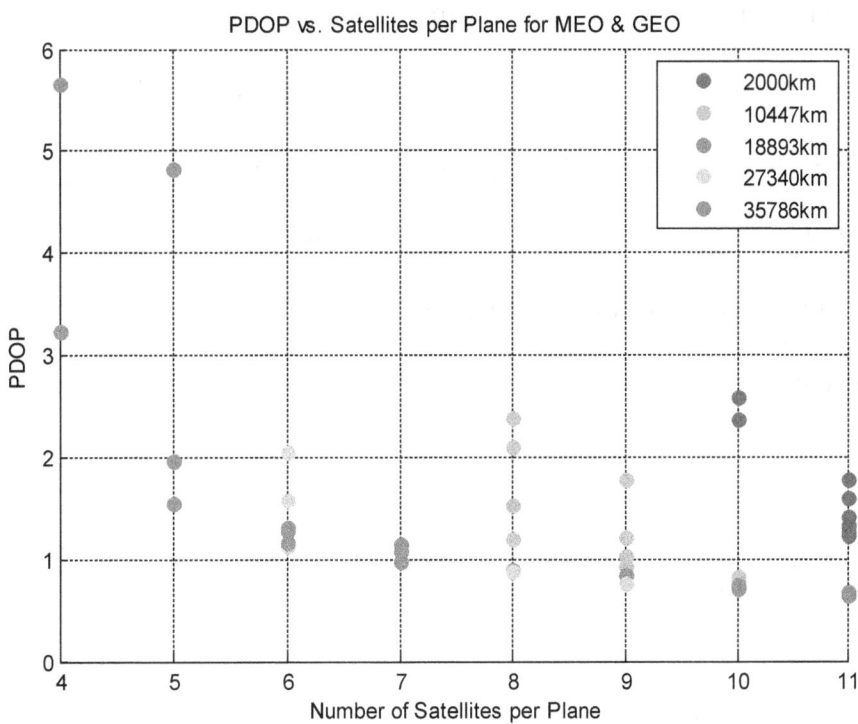

Figure 4-19: PDOP vs. Satellites per Plane for MEO & GEO

Figure 4-20 shows the relationship between the number of satellites per plane and the number of planes for the MEO and GEO test cases. For each altitude, as the number of planes increases, the number of satellites per plane either remains constant or increases. The largest difference between these variables is six, but this only occurred once. For the majority of points on this graph, the difference between the variables is less than six. The first and second MEO altitudes do not illustrate as much of an increase in satellites per plane as the number of planes increase. This could be attributed to the lower altitudes requiring more satellites in each plane to produce PDOP values less than six. With the other altitudes an increase in satellite per plane is more evident. The variety in satellites per plane for the upper altitudes could be a result of the higher altitudes providing better coverage with each orbital plane. Therefore, the number of satellites per plane does not remain at the upper bound values.

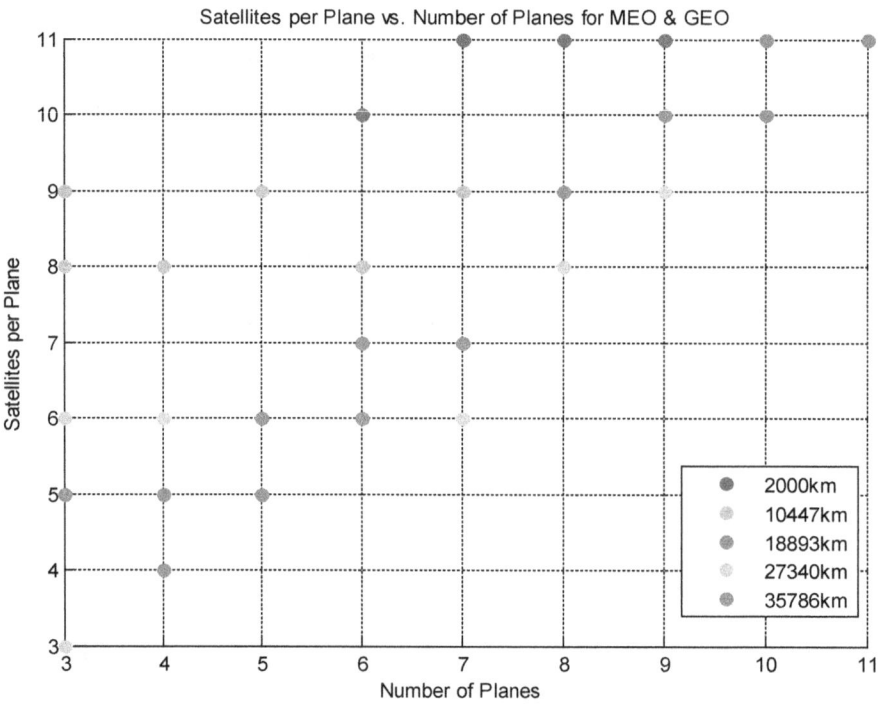

Figure 4-20: Satellites per Plane vs. Number of Planes for MEO & GEO

The second MEO test case, the dominating MEO test case, and the GEO test case illustrate a roughly linear trend between the number of satellites per plane and the number of planes. Equations 23 and 24 represent the respective trends between these parameters.

$$spp = 0.2471 * planes + 7.3817 \qquad (23)$$

Where R^2=0.83.

$$spp = 0.7623 * planes + 1.827 \qquad (24)$$

Where R^2=0.79.

In Figure 4-21 the inclination for MEO and GEO is compared to the PDOP values. There is no clear increase or decrease in PDOP as inclination changes. This is consistent with the results from the LEO altitudes. The first three altitudes demonstrate a 10 degree range of inclination, which matches the results from LEO. The upper MEO altitude and GEO did not illustrate the same range in inclination. Both altitudes were more sporadic, but stayed within a 20 degree range. Ignoring the upper MEO and GEO results, the two lower MEO altitudes were at a higher inclination range than the 18893 km results. A similar trend was seen with the LEO results. This makes sense that more inclination is needed at lower altitudes to produce PDOP values less than six.

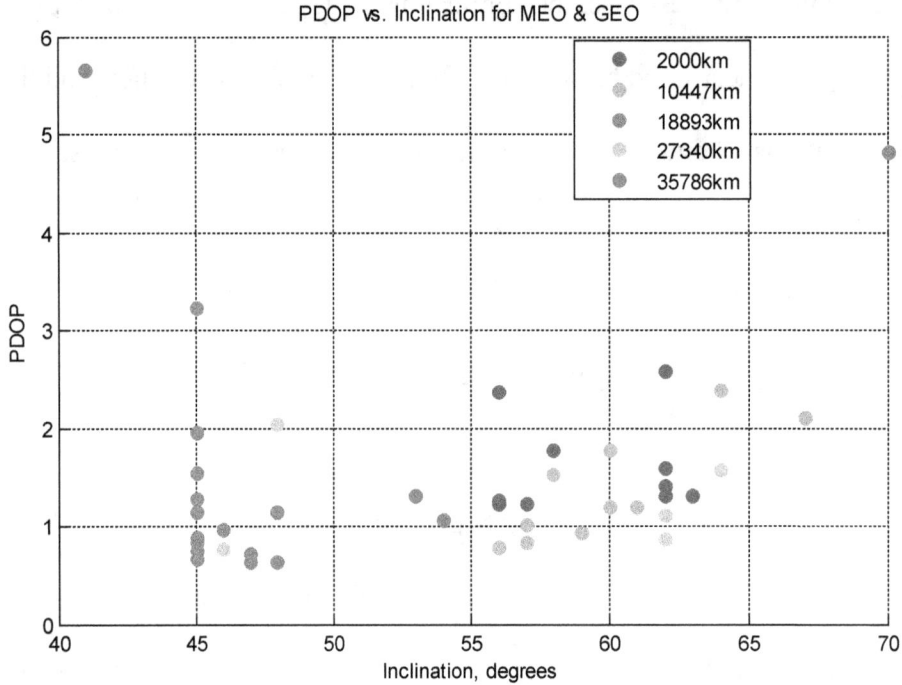

Figure 4-21: PDOP vs. Inclination for MEO & GEO

To analyze the effects on PDOP for solutions with the same number of planes and satellites per plane, Figure 4-22 shows the solutions from MEO and GEO with equal planes and satellites per plane. The upper MEO altitude and GEO altitude did not have any solutions with equal planes and satellites per plane. The results shown are from the first three MEO altitudes. For each set of solutions, the inclination is slightly higher for a lower PDOP value. This is consistent with the LEO results. Although inclination did not demonstrate a clear increase or decrease as PDOP changed for the overall solutions, the similar solutions illustrated that with a higher inclination it is possible to decrease PDOP.

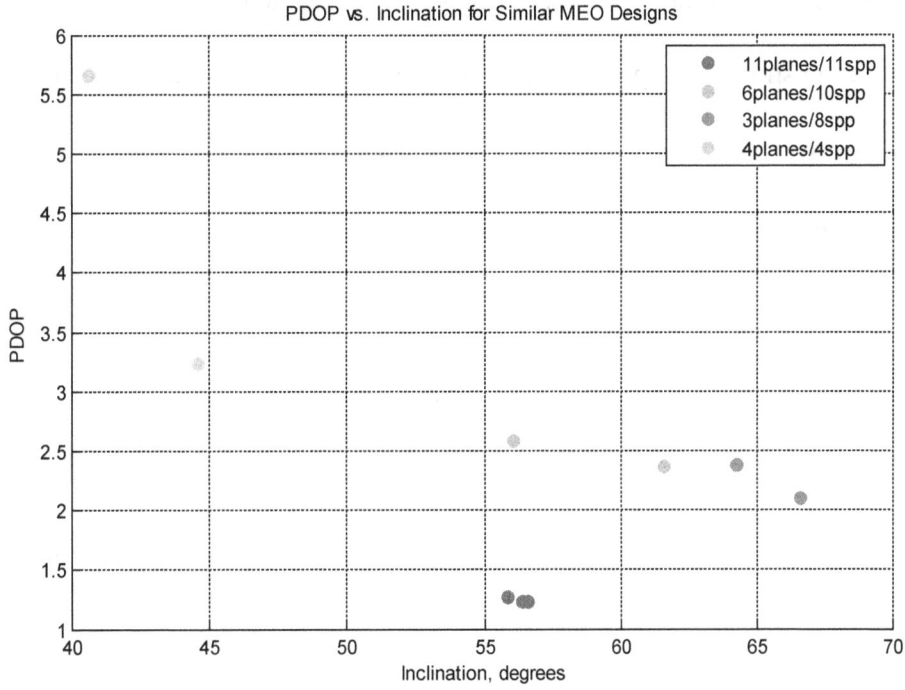

Figure 4-22: PDOP vs. Inclination for Similar MEO Designs

4.4 Analysis

The same designs used in Section 4.1.4 were compared to the values of the current GPS constellation. From each altitude range, the design from the concave portion of the respective Pareto fronts was used to analyze how PDOP changed with latitude. Figure 4-23 illustrates the PDOP values as latitude changes for both GPS and the best design in LEO. The GPS constellation possesses lower values for PDOP, and there are no major outliers. The LEO design possesses values for PDOP greater than six when latitude is less than -51 degrees or greater than 55 degrees. The LEO design shows increases in PDOP for latitudes less than -30 degrees or greater than 30 degrees. The GPS constellation shows the opposite trend for those latitudes as it decreases for latitudes less than -30 degrees or greater than 30 degrees. Since the GPS constellation is at a higher altitude, it is able to provide better geometry across the globe especially for higher latitudes.

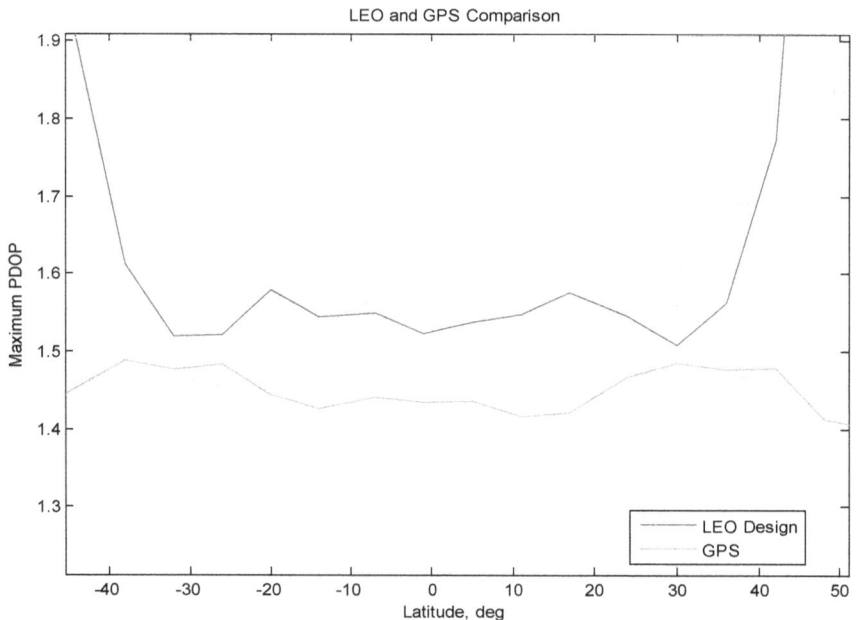

Figure 4-23: LEO and GPS Comparison

4-106

In Figure 4-24 the values of PDOP for the best MEO design are compared to the GPS as latitude changes. The MEO design produced lower values of PDOP for all latitudes. The MEO design was approximately 10,000 km lower than the GPS and possessed 21 more satellites. Both constellations show peaks from latitude of -55 degrees to -14 degrees and from 14 degrees to 55 degrees. This could be a result of both constellations having similar inclinations (around 55 degrees). The MEO design shows more of a decrease in PDOP around latitude of 0 degrees. The values decrease around zero degrees latitude because both constellations are inclined, so it is more difficult to have multiple satellites in view over the equator. For the validation case discussed in Section 3.5, the MOGA generated the current GPS constellation and not the constellation shown here. This is a result of a smaller range between the lower and upper bounds of the design parameters used for the GPS case.

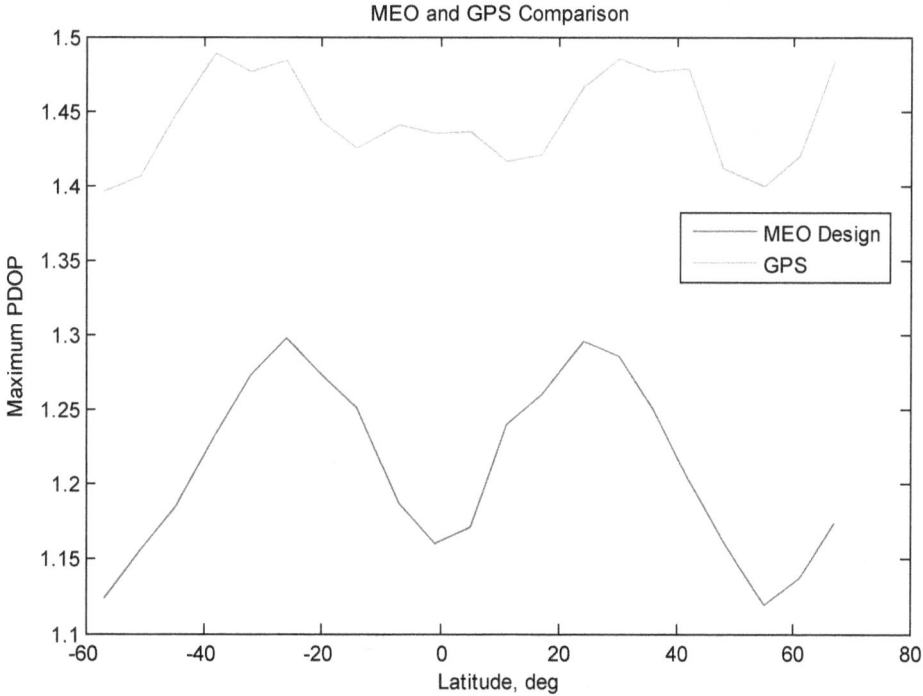

Figure 4-24: MEO and GPS Comparison

4-107

The GEO design produced lower PDOP values than GPS for all latitudes. Figure 4-25 illustrates the comparison of the two constellations. Both constellations show a drop in PDOP around latitude 48 degrees. For the GEO constellation, excluding latitudes less than -60 degrees or greater than 60 degrees, the highest values for PDOP occur around -7 to 5 degrees. However, for the GPS the PDOP values from latitude -7 to 5 degrees are lower than the others, excluding the same ranges as before. Both constellations maintain PDOP values less than six for all latitudes.

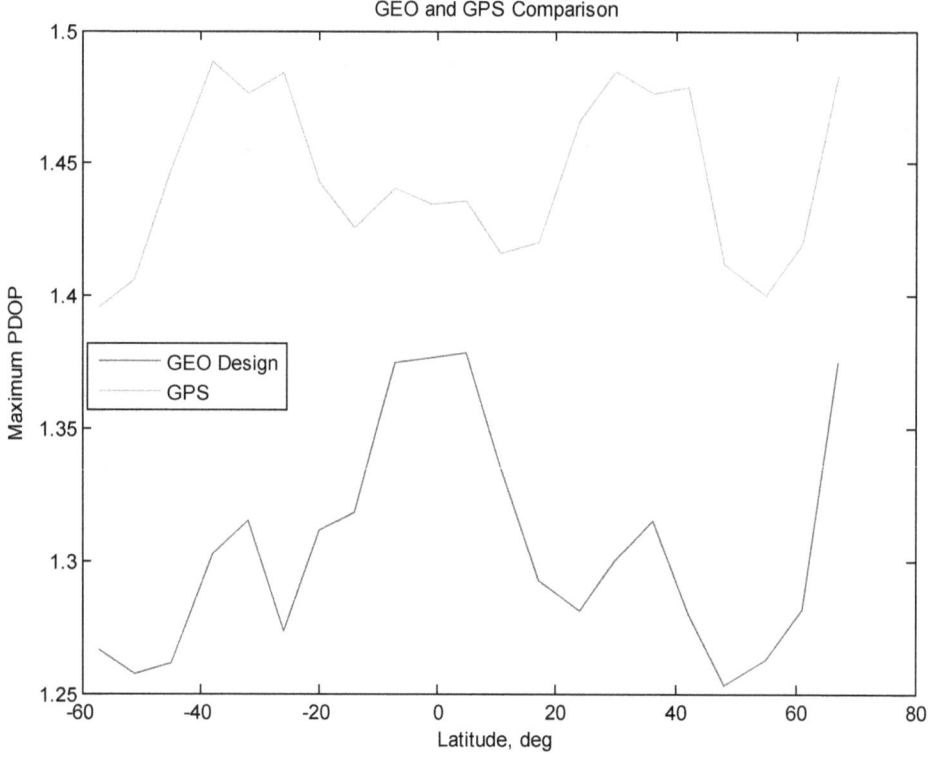

Figure 4-25: GEO and GPS Comparison

4.3 Refining Results

As a result of the method in which this problem was defined, the main advantage of utilizing the MOGA was in determining the RAAN increment, inclination, and the relationship between number of planes and satellites per plane. The altitude and transmit power were kept relatively constant, so the design space mostly consisted of the number of planes, satellites per plane, RAAN increment, and inclination. The number of planes and satellites per plane illustrated an obvious relationship between PDOP and cost. However, the inclination and RAAN increments were more random and did not illustrate an obvious relationship. As a result, it would be difficult to manually determine points on the Pareto front when varying these parameters. The MOGA would provide more of an advantage if the other design variables were analyzed with respect to this problem because the design space would increase.

For each altitude range, a design parameter sweep was completed for number of planes and satellites per plane. A point was selected on the Pareto front, and one of the parameters was held constant as the other one was varied. Figure 4-26 illustrates the parameter sweep for the number of planes using the LEO 3 test case. As the number of planes is varied, the points remain relatively close to the points on the Pareto front. The points generated from the parameters sweep did not result in a better PDOP for a lower cost.

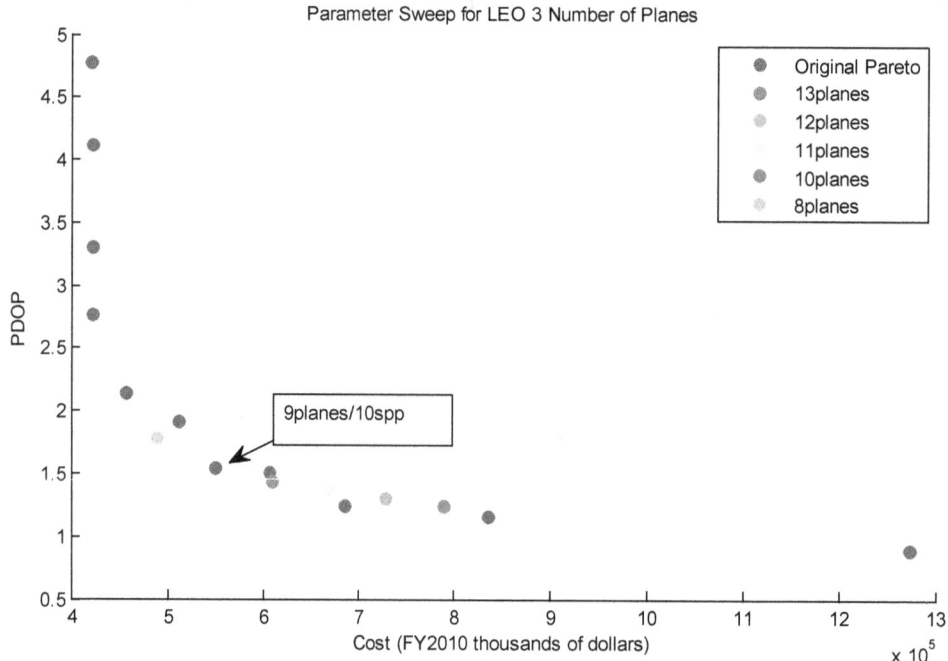

Figure 4-26: Parameter Sweep for LEO 3 Number of Planes

Figure 4-27 illustrates a parameter sweep of the satellites per plane in the LEO 3 test case. The points generated from the parameter sweep remain close to the points on the Pareto front. Both Figure 4-26 and Figure 4-27 show a strong relationship between the number of planes and satellites per plane. By varying these parameters, additional points were determined, but this is a result of using a small generation number for the simulations. If a larger generation number was used, it is possible the MOGA would determine these points as well.

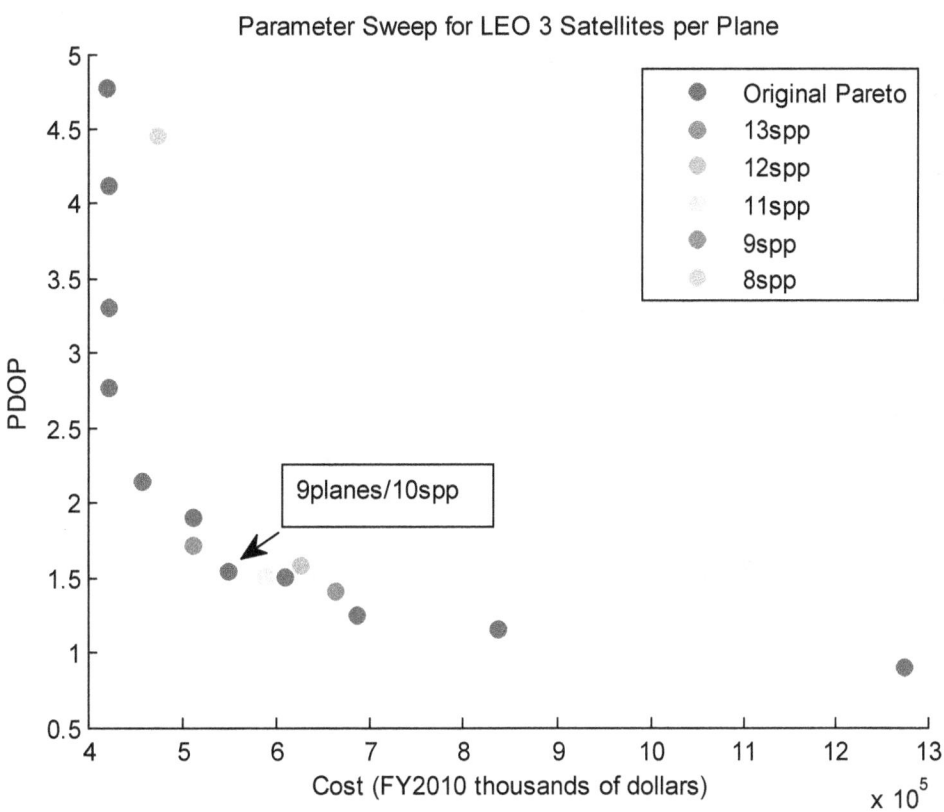

Figure 4-27: Parameter Sweep for LEO 3 Satellites per Plane

Figure 4-28 illustrates the sweep for the number of planes in the MEO 2 test case. These results are similar to those seen for the LEO altitude. Most of the points generated from the parameter sweep fell in between points on the Pareto front.

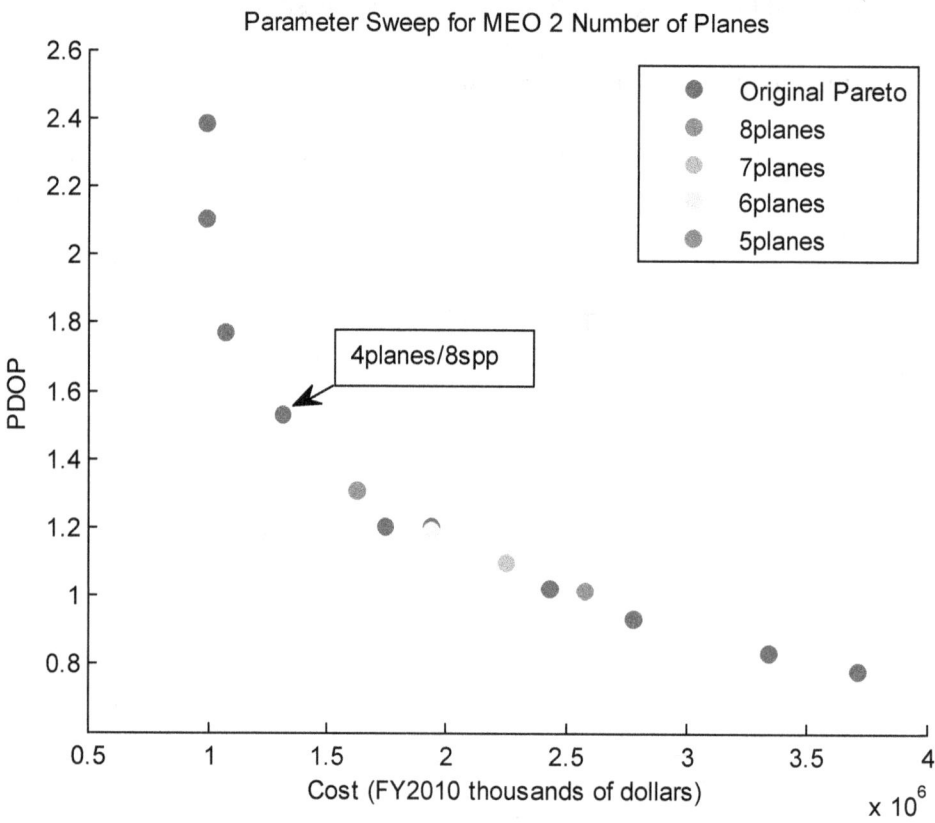

Figure 4-28: Parameter Sweep for MEO 2 Number of Planes

In Figure 4-29 the number of satellites per plane was varied for the MEO 2 test case. The parameter sweep produced a smooth curve just as seen with the previous results. The relationship between PDOP and cost remains the same as the number of satellites per plane decrease.

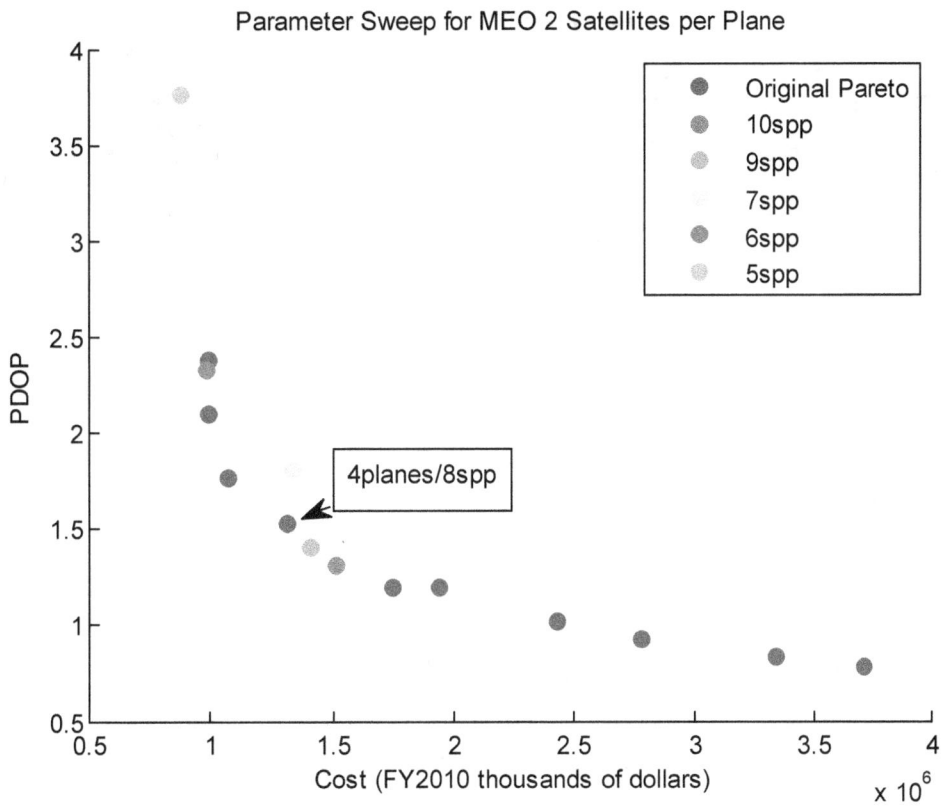

Figure 4-29: Parameter Sweep for MEO 2 Satellites per Plane

Figure 4-30 shows the sweep for the number of planes in the GEO test case. The parameter sweep generated points in between the points on the Pareto front. This is also a result of the low generation number used for the simulations. The point with six planes was almost exactly equal to one of the points on the Pareto front.

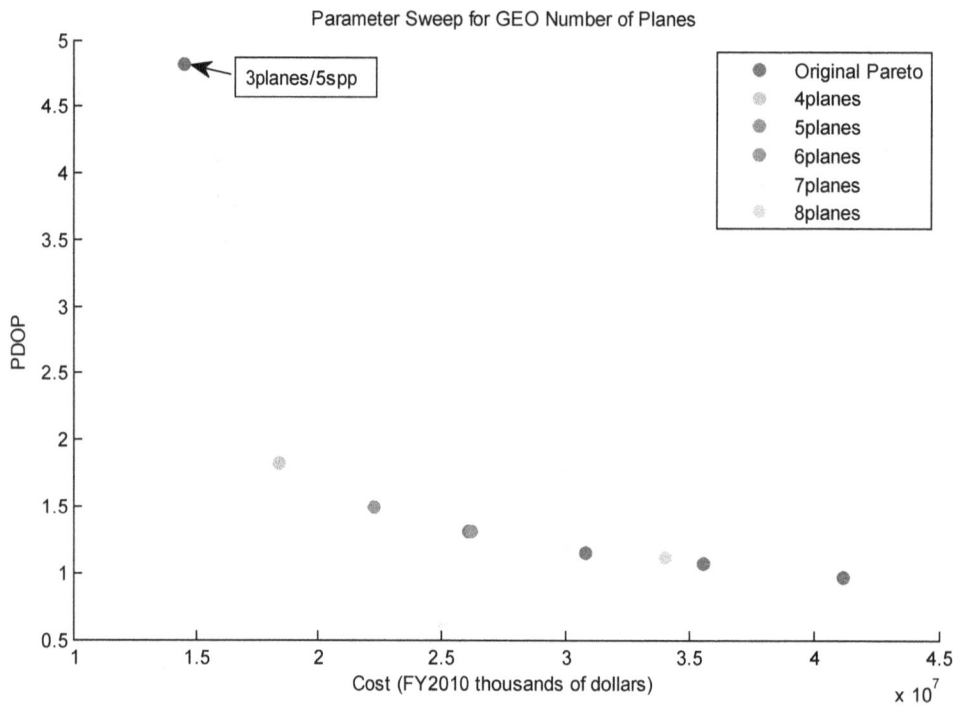

Figure 4-30: Parameter Sweep for GEO Number of Planes

Figure 4-31 illustrates the satellite per plane sweep for the GEO test case. These results are similar to those seen for the other two altitude ranges. All these parameters sweeps show that it is possible to determine points close to the Pareto front generated by the MOGA. None of the points generated from the parameter sweeps produced better PDOP values for a lower cost.

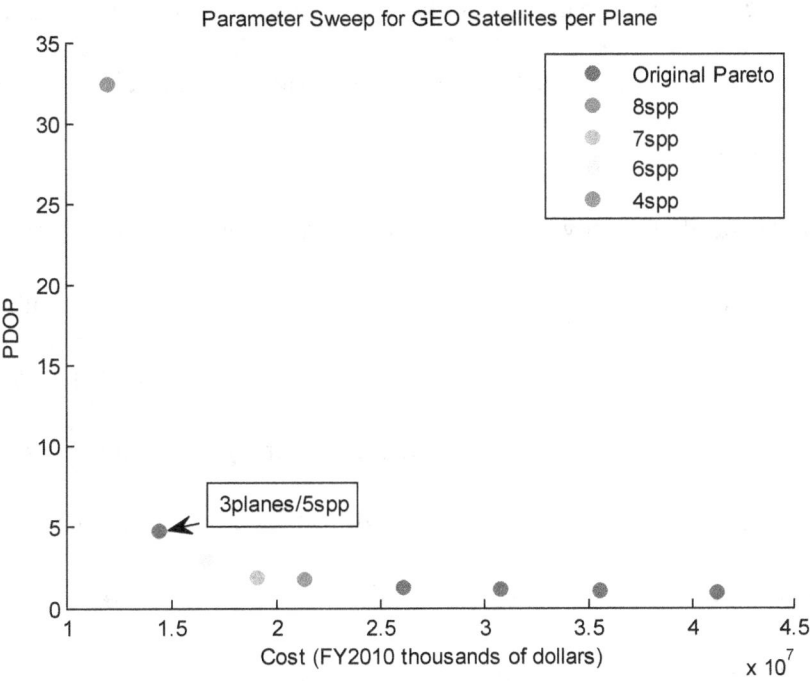

Figure 4-31: Parameter Sweep for GEO Satellites per Plane

Utilizing the MOGA for this problem was beneficial for determining the relationship between the number of planes and the number of satellites per plane. The MOGA was also beneficial in understanding the range of inclinations that are associated with the different orbital altitudes. Understanding these trends will allow for further research into navigation constellation design. Whether it is increasing the design space and continuing to use the MOGA for optimization or using a different optimization method, the trends and relationships determined from these results will be valuable.

4.5 Limitations

The results of this thesis demonstrate the use of MATLAB's MOGA and STK to produce and analyze navigation constellation designs. There are several limitations to the results of this research. There were eleven design parameters used in this analysis, but the eccentricity, argument of perigee, RAAN, and mean anomaly were set to zero. Therefore, the effects of these parameters on a navigation system at different altitudes were not determined. The difference in navigation performance of circular orbits versus elliptical orbits was not considered. The median of the maximum PDOP values was used to determine the central tendency of PDOP values across the latitudes. Therefore, there was no distribution in the PDOP values that were input to the MOGA. Another threshold could be easily implemented, but extensive sweeps were outside the scope of this work. The variation in PDOP over latitude was shown for several constellation designs, but the specific trend in PDOP was not analyzed for all altitudes included in this thesis.

Transmit power was calculated specifically for each altitude to ensure the proper size satellite and link closure. Therefore, the MOGA did not have the option of selecting cheaper satellites for orbits at a higher altitude. The spacecraft cost increased as the altitude increased to account for the transmit power required. This limited the spacecraft design options for constellations at higher altitudes. Walker constellations were used in this thesis, so all the satellites in a constellation were the same size. As a result, the performance and cost for constellations with mixed size satellites were not determined.

4.6 Summary

This chapter presented possible navigation constellation designs at different orbital altitudes. As expected, the PDOP values improved for designs at higher altitudes, and the number of satellites required decreased as altitude increased. The cost of the designs varied with altitude, but overall, the LEO designs illustrated less expensive options. Several of the designs were compared to the GPS constellation, and both the MEO and GEO designs maintained PDOP values lower than the GPS. The conclusions from these results are presented in the next chapter.

Chapter 5
Conclusions

The main contribution of this thesis was the multi-objective genetic algorithm model that ran in MATLAB in conjunction with STK and generated sets of navigation constellation designs. Through the generated sets of constellation designs, the tradeoffs between PDOP and system cost were illustrated. The results showed that the model used in this thesis was capable of creating realistic design solutions. The hybrid constellation results illustrated the ability for the design tool to handle multiple satellite types and multiple orbital altitudes.

5.1 Contributions

The design tool used in this thesis was developed, tested, and analyzed. This thesis accomplished the goal of creating a multi-objective design tool that could produce navigation constellation designs. The MOGA was able to search and check designs that a human might not try. The tradeoffs in the designs were analyzed, and the PDOP of several designs was evaluated over latitude. Several hybrid constellations were generated and compared to the individual altitude designs.

The navigation constellation designs generated in LEO possessed slightly higher PDOP values than MEO and GEO, but the constellations were less expensive. The results highlighted the interesting fact that even though the constellations in LEO required more satellites than MEO and GEO to produce PDOP values less than six, the cost was still less for those designs. The designs in LEO showed more changes in PDOP as altitude increased, but the cost was relatively similar. The designs in MEO illustrated the opposite effects.

PDOP values were analyzed over latitude for designs at MEO and GEO. They were then compared with PDOP over latitude for the current GPS constellation. Both MEO and GEO designs produced lower values for PDOP across the given latitudes. The MEO design also illustrated an increase in PDOP for the same latitudes as the GPS constellation.

The hybrid constellation results showed that by combining constellations, it is possible to achieve lower PDOP values. For the LEO-MEO hybrid constellation, there was little extra cost for the hybrid design, but for the MEO-GEO, there was a large cost increase compared to the MEO constellation. Most of the hybrid design solutions produced PDOP values less than one, which was much lower than many of the design solutions from the individual altitudes.

5.2 Challenges

One of the most challenging aspects of this research was the limitation in computation time. When developing the design tool used in this thesis, it was difficult to determine the origins of an error because of the model complexity. The algorithm would run for several hours before notifying the user of an error. When a change was made to the model, the user had to wait a couple days to see the effects. Due to the size of this model, the software programs, STK and MATLAB, would crash during several simulations. Possessing multiple computers to run simulations would add redundancy and prevent having to completely restart simulations if the programs crash.

Another challenge was attempting to complete STK tasks within MATLAB code. To execute STK tasks within MATLAB, multiple references were needed. The example

5-119

commands available in STK did not always provide the level of detail required to develop a functioning command. Determining the cause of an error was also challenging due to the lack of detail in the error message.

Using a multi-objective genetic algorithm illustrated the tradeoffs between the two objective functions used in this thesis, but it was difficult to develop the problem without the ability to use a constraint function. The constraint function could have been applied to the launch vehicle analysis. It could have also been used to force certain design variables to be integer values. Understanding the MATLAB code for the mutation and crossover functions is not trivial, and attempting to use those functions to constrain certain variables was difficult. The limitation in computation time made it difficult to ensure the updates to the mutation and crossover functions were working correctly. This challenge is the reason for creating test cases based on altitude and transmit power.

5.3 Recommendations for Future Work

This thesis demonstrated the ability to use MATLAB's MOGA along with STK to generate navigation constellation designs. This design tool could be adapted for other uses by changing the design vector and/or objective functions to meet the user's requirements. The design tool used in this thesis was able to exploit high levels of fidelity from STK. STK's extensive library of evaluations could easily be changed for many complicated objectives. The design tool used in this thesis could be expanded and further explored.

This thesis included launch vehicle cost, but it did not explore the effects of bundling the spacecraft together on launch vehicles in various configurations. The launch vehicle cost used in this thesis was determined by assuming direct launches into the design orbits. An analysis of inserting a spacecraft directly into an orbit compared to utilizing a transfer orbit could give details on a more optimal method for creating the constellations generated from the design tool.

This thesis focused on the navigation metric, PDOP, but there are other metrics for navigation that could be examined. PDOP is a geometric measure, but STK also offers the ability to analyze actual position error from the measurements. This could illustrate the level of accuracy available on the ground when using new constellation designs based on clock and orbit uncertainty. It could also illustrate areas across the globe where accuracy will be lower and require additional assets to achieve reasonable positioning.

This thesis focused on the global value for PDOP, but it did not focus specifically on certain target areas. Future research could narrow the analysis to a few targets to analyze the change in accuracy over time or how certain regions compared to one another in terms of PDOP values. The values for PDOP at the specified targets could be analyzed with different constellation designs to determine if one design offers better PDOP values at latitudes where it is difficult to achieve accurate positioning.

This thesis exclusively utilized MATLAB's MOGA to optimize PDOP and cost. Future work may involve exploring different optimization algorithms. Different optimizers were discussed briefly in Section 2.4. Further improvements could be made with the MOGA as well. Using a higher population size or generation number would generate more accurate results. Improving the computation time for the simulations would allow for more research into the MOGA itself.

The hybrid constellation designs in this thesis were composed of two different constellations. Additional research could be done in analyzing these designs. Due to the computational limitation, a sufficient amount of research was unable to be accomplished with the hybrid constellations. Future work could examine the appropriate bounds to use for the design parameters and attempt to improve the resolution of the Pareto fronts for those constellations.

Although there is additional research that could be accomplished to improve the design tool developed in this research, this thesis provided additional information on using MATLAB's MOGA along with STK to generate navigation constellation designs. This research illustrated the advantages and disadvantages of using this method, which overall provided further insight into using this constellation design tool specifically for the use of navigation.

www.ingramcontent.com/pod-product-compliance
Lightning Source LLC
Chambersburg PA
CBHW081154180526
45170CB00006B/2070